SAXON MATH™ 2

Student Workbook

Part 2

Nancy Larson

with Roseann Paolino
Maureen Hannan

A Harcourt Achieve Imprint

www.SaxonPublishers.com
1-800-284-7019

ISBN-13: 978-1-6003-2721-6 (Set)
ISBN-10: 1-6003-2721-4 (Set)
ISBN-13: 978-1-6003-2574-8 (Part 1)
ISBN-10: 1-6003-2574-2 (Part 1)
ISBN-13: 978-1-6003-2577-9 (Part 2)
ISBN-10: 1-6003-2577-7 (Part 2)

Printed in the United States of America

5 6 7 8 9 0982 14 13 12 11 10 09

Name ______________________________

Class Fact Practice **71A**

***Saxon** Math 2 (for use with **Lesson 71**)*

Set 14: Subtracting 2, 1, and 0

Do −2 Wrap-Up once.
Do −1 Wrap-Up once.

$8 - 2$	$5 - 2$	$10 - 2$	$3 - 2$	$9 - 2$
$6 - 2$	$11 - 2$	$2 - 2$	$7 - 2$	$4 - 2$
$8 - 1$	$10 - 1$	$4 - 1$	$6 - 1$	$1 - 1$
$9 - 1$	$3 - 1$	$7 - 1$	$2 - 1$	$5 - 1$
$5 - 0$	$6 - 0$	$3 - 0$	$8 - 0$	$1 - 0$

Name ______________________________

Fact Homework 71B

Saxon *Math 2 (for use with* ***Lesson 71****)*

Set 14: Subtracting 2, 1, and 0

1. Read the answers to someone.
2. Write the answers.
3. Ask someone to correct your paper. Corrected by ______________________

$$\begin{array}{r}5\\-\ 2\\\hline\end{array}\quad\begin{array}{r}10\\-\ 2\\\hline\end{array}\quad\begin{array}{r}3\\-\ 2\\\hline\end{array}\quad\begin{array}{r}9\\-\ 2\\\hline\end{array}\quad\begin{array}{r}7\\-\ 2\\\hline\end{array}$$

$$\begin{array}{r}11\\-\ 2\\\hline\end{array}\quad\begin{array}{r}4\\-\ 2\\\hline\end{array}\quad\begin{array}{r}8\\-\ 2\\\hline\end{array}\quad\begin{array}{r}6\\-\ 2\\\hline\end{array}\quad\begin{array}{r}2\\-\ 2\\\hline\end{array}$$

$$\begin{array}{r}3\\-\ 1\\\hline\end{array}\quad\begin{array}{r}9\\-\ 1\\\hline\end{array}\quad\begin{array}{r}5\\-\ 1\\\hline\end{array}\quad\begin{array}{r}1\\-\ 1\\\hline\end{array}\quad\begin{array}{r}7\\-\ 1\\\hline\end{array}$$

$$\begin{array}{r}10\\-\ 1\\\hline\end{array}\quad\begin{array}{r}6\\-\ 1\\\hline\end{array}\quad\begin{array}{r}2\\-\ 1\\\hline\end{array}\quad\begin{array}{r}8\\-\ 1\\\hline\end{array}\quad\begin{array}{r}4\\-\ 1\\\hline\end{array}$$

$$\begin{array}{r}7\\-\ 0\\\hline\end{array}\quad\begin{array}{r}0\\-\ 0\\\hline\end{array}\quad\begin{array}{r}9\\-\ 0\\\hline\end{array}\quad\begin{array}{r}4\\-\ 0\\\hline\end{array}\quad\begin{array}{r}2\\-\ 0\\\hline\end{array}$$

Name ____________________

Date ____________________

Guided Class Practice 71A

***Saxon** Math 2 (for use with **Lesson 71**)*

1. Curtis had 65¢. His sister gave him 27¢. How much money does he have now?

Number sentence ____________________

Answer ____________________

Workspace

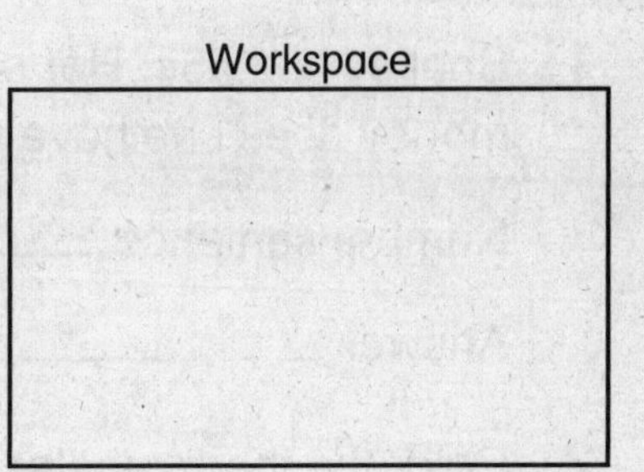

2. Color the thermometer to show 38°F.

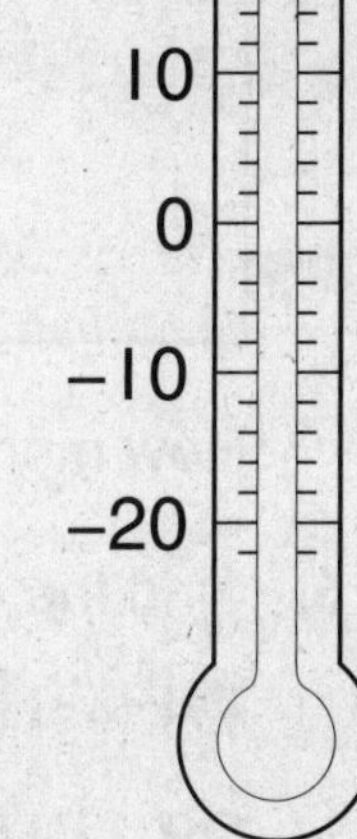

3. Write the fraction that tells how much is shaded. ___

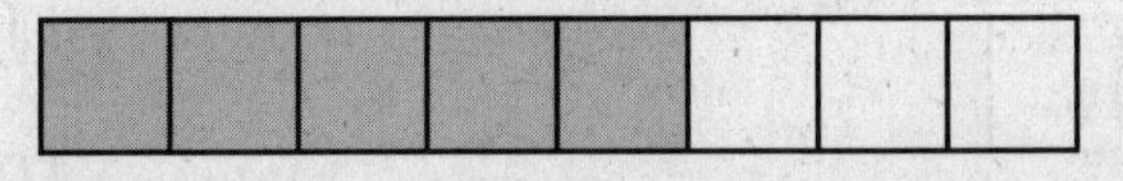

Write the fraction that tells how much is not shaded. ___

4. Fill in the children's names to show the ice cream flavors they like.

Amy likes only chocolate.
Chris likes vanilla and chocolate.
Sue likes only vanilla.
Bob likes only vanilla.
Jim likes both flavors.

How many children like chocolate? ___

How many children like vanilla? ___

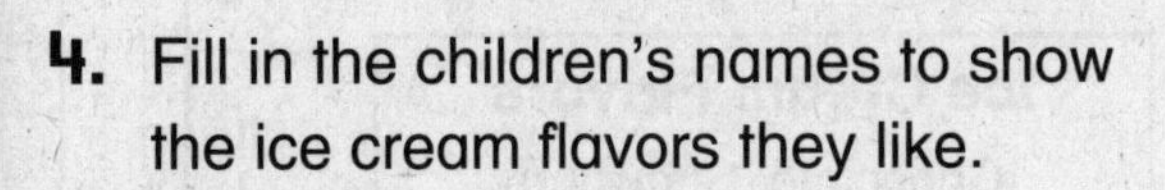

5. I have 3 dimes, 2 nickels, and 4 pennies. Draw the coins.

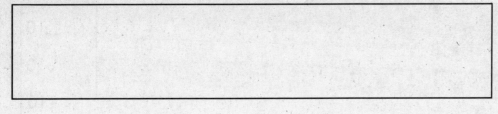

How much money is this? ___

6. Find the answers.

67 − 10 = ___

43 − 10 = ___

19 − 10 = ___

13¢	15¢
42¢	23¢
+ 38¢	+ 37¢

Name ________________________

Date ________________________

Workspace

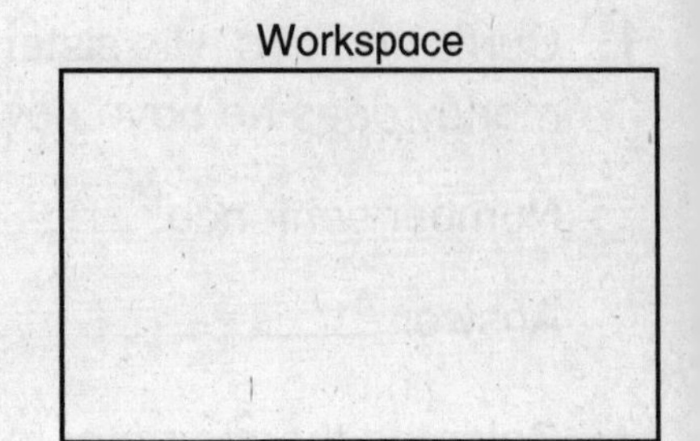

1. Shelley had 55¢. Her sister gave her 38¢. How much money does she have now?

Number sentence ________________

Answer ________________

2. Color the thermometer to show 54°F.

°F

110 100 90 80 70 60 50 40 30 20 10 0 −10 −20

3. Write the fraction that tells how much is shaded. ___

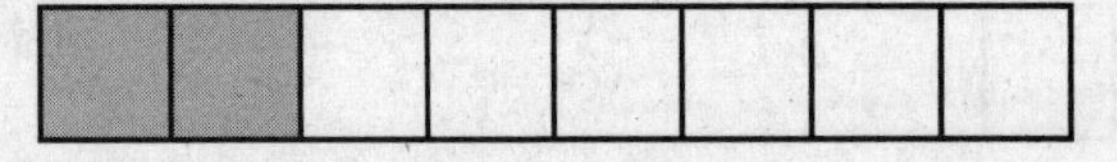

Write the fraction that tells how much is not shaded. ___

4. Fill in the children's names to show the ice cream flavors they like.

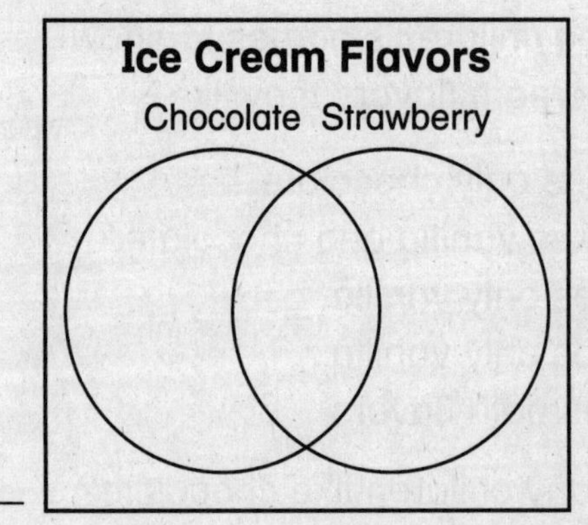

Mike likes only strawberry.
Steve likes both flavors.
Jim likes only chocolate.
Mary likes both flavors.
Lisa likes only strawberry.

How many children like chocolate? ______

How many children like strawberry? ______

5. I have 5 dimes, 3 nickels, and 1 penny. Draw the coins.

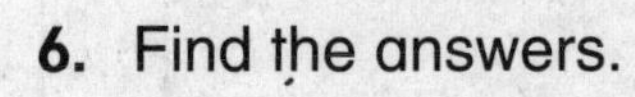

How much money is this? ______

6. Find the answers.

74 − 10 = ______

28 − 10 = ______

16 − 10 = ______

35¢	14¢
13¢	45¢
+ 32¢	+ 26¢

Name ____________________

Class Fact Practice **72A**

Saxon *Math 2 (for use with* ***Lesson 72****)*

Set 14: Subtracting 2, 1, and 0

Do –2 Wrap-Up once. ☐ Do –1 Wrap-Up once. ☐ Do –0 Wrap-Up once. ☐

3 – 2 = ____	5 – 1 = ____	6 – 0 = ____
9 – 2 = ____	2 – 1 = ____	0 – 0 = ____
5 – 2 = ____	10 – 1 = ____	8 – 0 = ____
10 – 2 = ____	7 – 1 = ____	4 – 0 = ____
2 – 2 = ____	3 – 1 = ____	7 – 0 = ____
8 – 2 = ____	8 – 1 = ____	1 – 0 = ____
6 – 2 = ____	4 – 1 = ____	9 – 0 = ____
11 – 2 = ____	1 – 1 = ____	2 – 0 = ____
7 – 2 = ____	9 – 1 = ____	5 – 0 = ____
4 – 2 = ____	6 – 1 = ____	3 – 0 = ____

Name ____________________

Saxon Math 2 (for use with Lesson 72)

Set 14: Subtracting 2, 1, and 0

1. Read the answers to someone.
2. Write the answers.
3. Ask someone to correct your paper. Corrected by ____________________

$5 - 2 =$ ____	$10 - 1 =$ ____	$8 - 2 =$ ____
$6 - 1 =$ ____	$2 - 2 =$ ____	$1 - 1 =$ ____
$7 - 2 =$ ____	$8 - 0 =$ ____	$11 - 2 =$ ____
$3 - 2 =$ ____	$9 - 1 =$ ____	$0 - 0 =$ ____
$4 - 1 =$ ____	$6 - 2 =$ ____	$7 - 1 =$ ____
$10 - 2 =$ ____	$5 - 0 =$ ____	$4 - 2 =$ ____
$5 - 1 =$ ____	$9 - 2 =$ ____	$3 - 1 =$ ____
$7 - 0 =$ ____	$2 - 1 =$ ____	$8 - 1 =$ ____

Name ___________________________________

Lesson Worksheet 72

***Saxon** Math 2 (for use with **Lesson 72**)*

Draw these line segments.

3" •

$2\frac{1}{2}$" •

$1\frac{1}{2}$" •

$\frac{1}{2}$" •

$4\frac{1}{2}$" •

Measure these line segments.

☐ **1.** •————————————————•

☐ **2.** •——————————————————————————•

☐ **3.** •———————•

☐ **4.** •————————————•

☐ **5.** •——•

Name ______________________

Date ______________________

Guided Class Practice 72A

***Saxon** Math 2 (for use with **Lesson 72**)*

1. Theresa has 4 dimes and 7 pennies.
How much money does Theresa have? ________

Jennifer has 2 dimes and 5 pennies.
How much money does Jennifer have? ________

How much money do the two girls have altogether?

Number sentence ______________________

Answer ______________________

Workspace

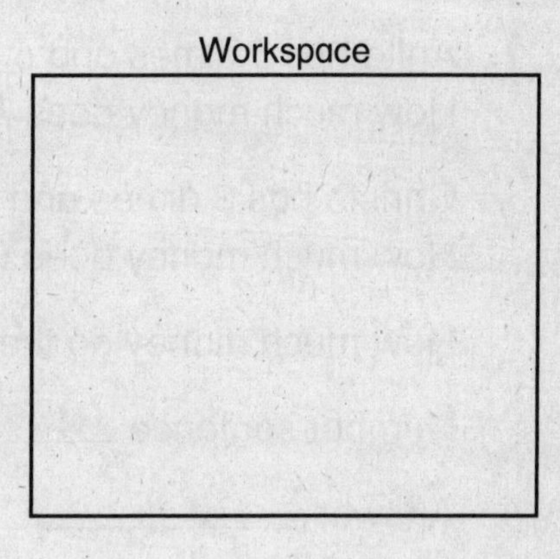

2. Draw one dozen eggs.
Color a half dozen red.
What fractional part of the eggs are not red?

Answer ——

3. Color the shapes that are congruent to the shape on the left.

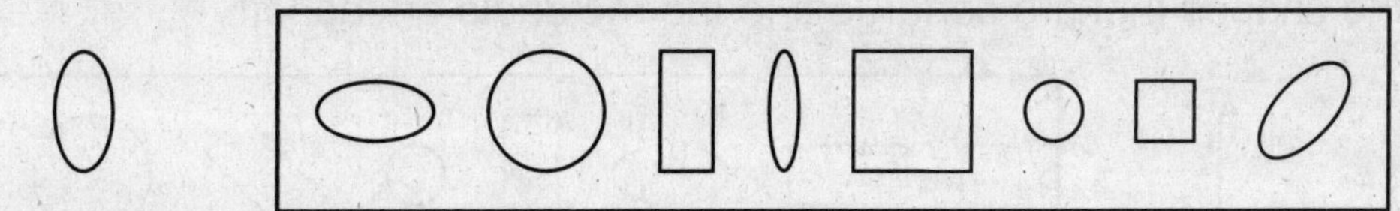

4. Find the answers.

34 − 10 = ______ 62 − 10 = ______ 17 − 10 = ______

5. Use your ruler to draw these line segments.

$5\frac{1}{2}$" •

$\frac{1}{2}$" •

6. Write the problems vertically. Find the sums.

63¢ + 9¢

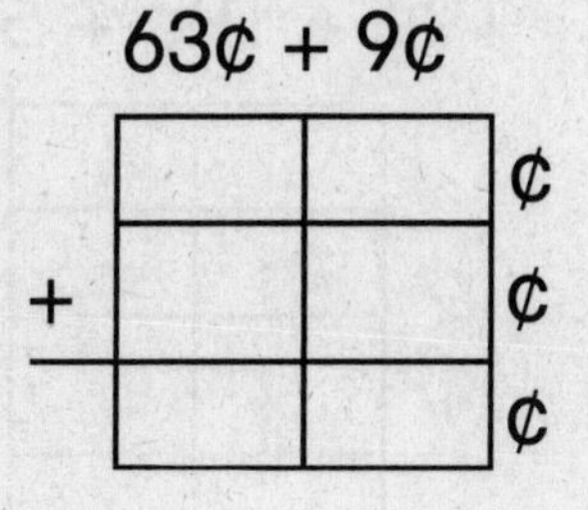

27¢ + 53¢

54¢ + 12¢ + 24¢

Name ______________________

Date ______________________

Homework 72B

***Saxon** Math 2 (for use with **Lesson 72**)*

1. Anita has 5 dimes and 6 pennies.
How much money does Anita have? __________

Christa has 3 dimes and 4 pennies.
How much money does Christa have? __________

How much money do the two girls have altogether?

Number sentence ______________________

Answer ______________________

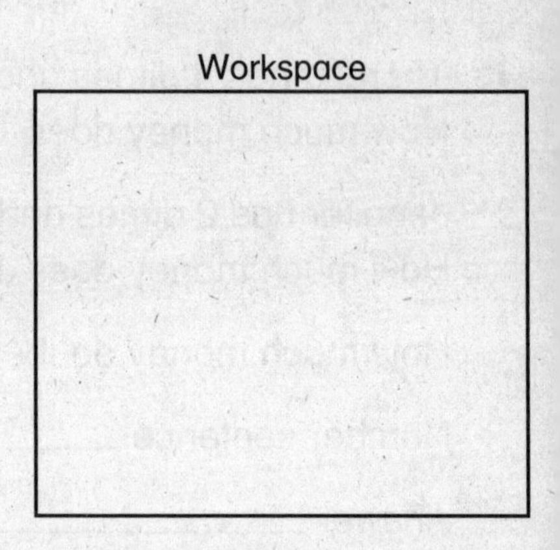

2. Draw one dozen oranges.
Color a half dozen orange.
What fractional part of the oranges are not colored?

Answer _____

3. Color the shapes that are congruent to the rectangle on the left.

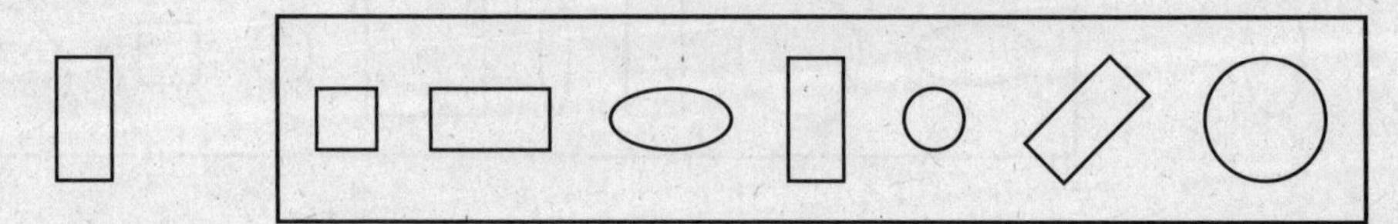

4. Find the answers.

59 − 10 = ______ 18 − 10 = ______ 75 − 10 = ______

5. Finish these number patterns.

20, 22, 24, 26, ______, ______, ______, ______, ______

50, 45, 40, 35, ______, ______, ______, ______, ______

6. Write the problems vertically. Find the sums.

54¢ + 16¢

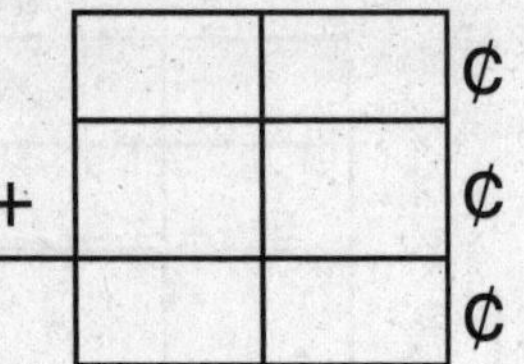

63¢ + 24¢

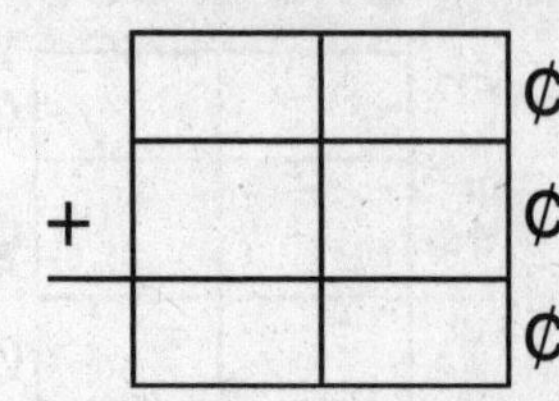

19¢ + 14¢ + 51¢

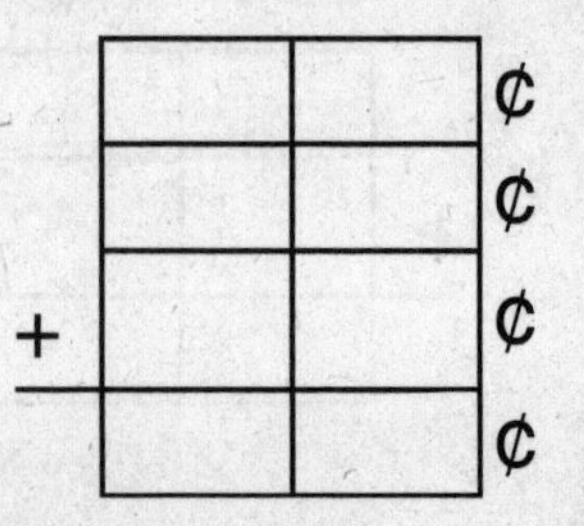

Name ______________________________

Class Fact Practice 73A

***Saxon** Math 2 (for use with **Lesson 73**)*

Set 14: Subtracting 2, 1, and 0

$6 - 2$	$9 - 1$	$7 - 0$	$10 - 2$	$3 - 1$
$7 - 2$	$2 - 1$	$3 - 0$	$5 - 1$	$3 - 2$
$8 - 1$	$5 - 2$	$1 - 0$	$9 - 2$	$6 - 1$
$11 - 2$	$4 - 1$	$2 - 2$	$10 - 1$	$8 - 0$
$7 - 1$	$4 - 2$	$5 - 0$	$8 - 2$	$1 - 1$

Name ______________________________

Saxon Math 2 (for use with Lesson 73)

Set 14: Subtracting 2, 1, and 0

1. Read the answers to someone.
2. Write the answers.
3. Ask someone to correct your paper. Corrected by ____________________

3 − 2	5 − 1	3 − 0	2 − 1	7 − 2
3 − 1	10 − 2	7 − 0	9 − 1	6 − 2
1 − 1	8 − 2	5 − 0	4 − 2	7 − 1
8 − 0	10 − 1	2 − 2	4 − 1	11 − 2
6 − 1	9 − 2	1 − 0	5 − 2	8 − 1

Name ______________________________

Date ______________________________

Guided Class Practice 73A

Saxon Math 2 (for use with Lesson 73)

1. Fifteen Grade 2 children rode bicycles to school. Seventeen Grade 3 children rode bicycles to school. How many children in Grades 2 and 3 rode bicycles to school?

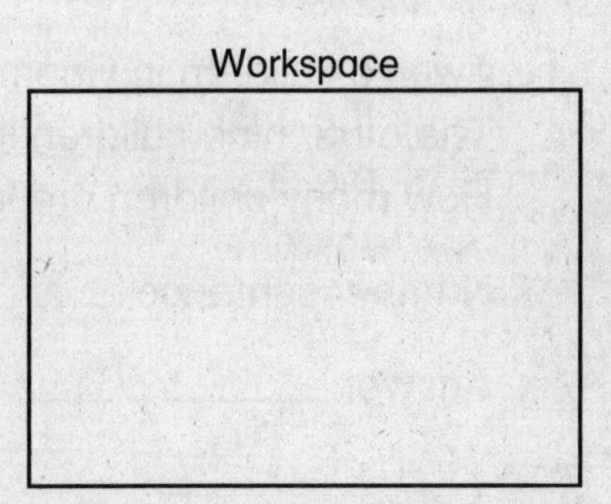

Number sentence ______________________

Answer ______________________

2. What is the temperature on the thermometer? ______ °F

°F
110
100
90
80
70
60
50
40
30
20
10
0
−10
−20

3. Use your ruler to measure these line segments.

•——————————• ______ "

•————————————————————————————• ______ "

4. Divide each shape into fourths.

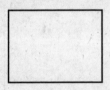

Shade $\frac{3}{4}$.

Shade $\frac{2}{4}$.

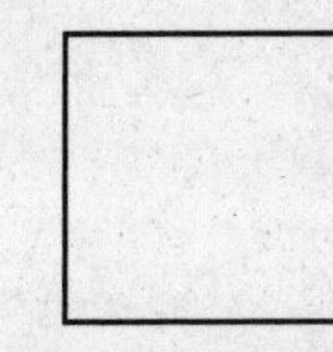

Shade $\frac{1}{4}$.

5. Use the graph to answer the questions.

How many children like hiking? ______

How many children like skating? ______

How many children like only hiking? ______

What does Curt like? ______________________

____

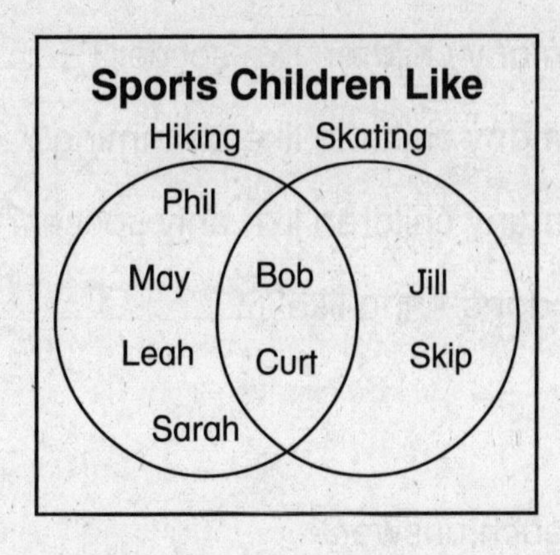

6. Find each answer.

$51 - 10 =$ ______

$65 - 10 =$ ______

one more than $26 =$ ______

ten more than $28 =$ ______

$$\begin{array}{r} 58 \\ +\ 59 \\ \hline \end{array}$$

Name ______________________

Date ______________________

Homework 73B

***Saxon** Math 2 (for use with **Lesson 73**)*

1. Twelve children in Room 16 ride the bus to school. The other nine children in the class walk to school. How many children are in Room 16?

Number sentence ______________________

Answer ______________________

Workspace

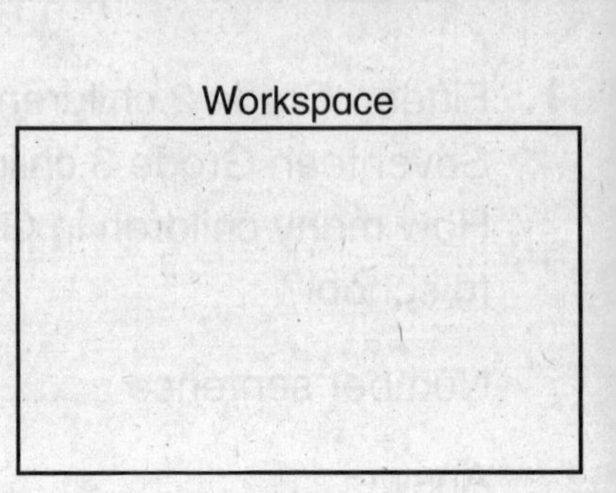

2. What is the temperature on the thermometer? ________ °F

°F 110 100 90 80 70 60 50 40 30 20 10 0 −10 −20

3. What is the best estimate of the length of this paper?

3 inches   28 inches   11 inches   20 inches

4. Divide each shape into fourths.

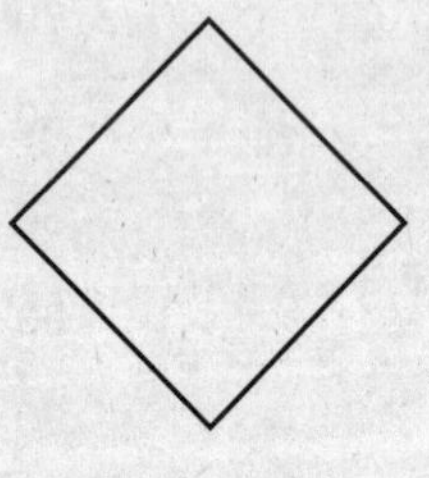

Shade $\frac{1}{4}$.

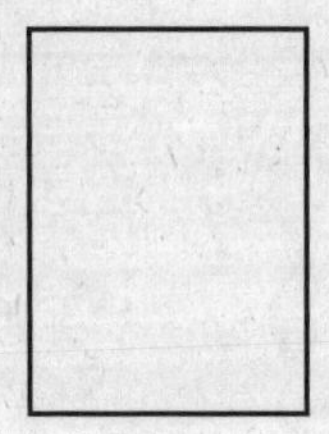

Shade $\frac{2}{4}$.

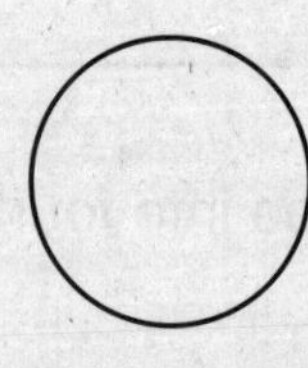

Shade $\frac{3}{4}$.

5. Use the graph to answer the questions.

How many children like soccer? ________

How many children like swimming? ________

How many children like only soccer? ________

What does Anna like? ______________

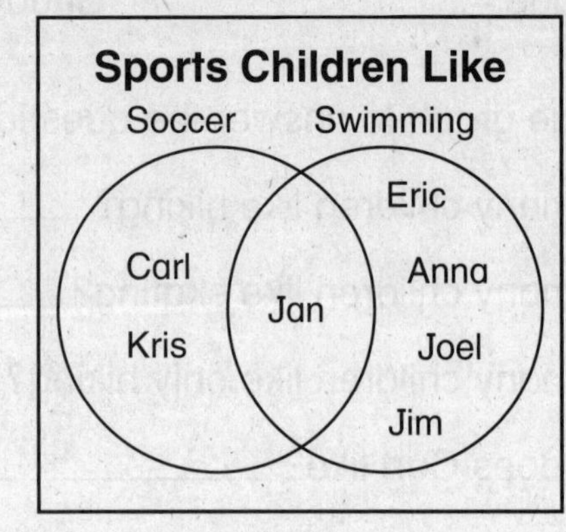

6. Find each answer.

$64 - 10 =$ ________   ten more than $61 =$ ________

$39 - 10 =$ ________   ten less than $40 =$ ________

$$\begin{array}{r} 72 \\ +\ 48 \\ \hline \end{array}$$

Name ______________________________

Set 14: Subtracting 2, 1, and 0

Do –2 Wrap-Up twice.
Do –1 Wrap-Up once.

$$\begin{array}{r} 8 \\ -\ 0 \\ \hline \end{array} \quad \begin{array}{r} 10 \\ -\ 1 \\ \hline \end{array} \quad \begin{array}{r} 2 \\ -\ 2 \\ \hline \end{array} \quad \begin{array}{r} 4 \\ -\ 1 \\ \hline \end{array} \quad \begin{array}{r} 11 \\ -\ 2 \\ \hline \end{array}$$

$$\begin{array}{r} 1 \\ -\ 1 \\ \hline \end{array} \quad \begin{array}{r} 8 \\ -\ 2 \\ \hline \end{array} \quad \begin{array}{r} 5 \\ -\ 0 \\ \hline \end{array} \quad \begin{array}{r} 4 \\ -\ 2 \\ \hline \end{array} \quad \begin{array}{r} 7 \\ -\ 1 \\ \hline \end{array}$$

$$\begin{array}{r} 3 \\ -\ 1 \\ \hline \end{array} \quad \begin{array}{r} 10 \\ -\ 2 \\ \hline \end{array} \quad \begin{array}{r} 7 \\ -\ 0 \\ \hline \end{array} \quad \begin{array}{r} 9 \\ -\ 1 \\ \hline \end{array} \quad \begin{array}{r} 6 \\ -\ 2 \\ \hline \end{array}$$

$$\begin{array}{r} 6 \\ -\ 1 \\ \hline \end{array} \quad \begin{array}{r} 9 \\ -\ 2 \\ \hline \end{array} \quad \begin{array}{r} 1 \\ -\ 0 \\ \hline \end{array} \quad \begin{array}{r} 5 \\ -\ 2 \\ \hline \end{array} \quad \begin{array}{r} 8 \\ -\ 1 \\ \hline \end{array}$$

$$\begin{array}{r} 3 \\ -\ 2 \\ \hline \end{array} \quad \begin{array}{r} 5 \\ -\ 1 \\ \hline \end{array} \quad \begin{array}{r} 3 \\ -\ 0 \\ \hline \end{array} \quad \begin{array}{r} 2 \\ -\ 1 \\ \hline \end{array} \quad \begin{array}{r} 7 \\ -\ 2 \\ \hline \end{array}$$

Name ______________________________

Fact Homework **74B**

***Saxon** Math 2 (for use with **Lesson 74**)*

Set 14: Subtracting 2, 1, and 0

1. Read the answers to someone.
2. Write the answers.
3. Ask someone to correct your paper. Corrected by ______________________

$\begin{array}{r} 7 \\ -\ 2 \\ \hline \end{array}$	$\begin{array}{r} 2 \\ -\ 1 \\ \hline \end{array}$	$\begin{array}{r} 3 \\ -\ 0 \\ \hline \end{array}$	$\begin{array}{r} 5 \\ -\ 1 \\ \hline \end{array}$	$\begin{array}{r} 3 \\ -\ 2 \\ \hline \end{array}$
$\begin{array}{r} 8 \\ -\ 1 \\ \hline \end{array}$	$\begin{array}{r} 5 \\ -\ 2 \\ \hline \end{array}$	$\begin{array}{r} 1 \\ -\ 0 \\ \hline \end{array}$	$\begin{array}{r} 9 \\ -\ 2 \\ \hline \end{array}$	$\begin{array}{r} 6 \\ -\ 1 \\ \hline \end{array}$
$\begin{array}{r} 11 \\ -\ 2 \\ \hline \end{array}$	$\begin{array}{r} 4 \\ -\ 1 \\ \hline \end{array}$	$\begin{array}{r} 2 \\ -\ 2 \\ \hline \end{array}$	$\begin{array}{r} 10 \\ -\ 1 \\ \hline \end{array}$	$\begin{array}{r} 8 \\ -\ 0 \\ \hline \end{array}$
$\begin{array}{r} 7 \\ -\ 1 \\ \hline \end{array}$	$\begin{array}{r} 4 \\ -\ 2 \\ \hline \end{array}$	$\begin{array}{r} 5 \\ -\ 0 \\ \hline \end{array}$	$\begin{array}{r} 8 \\ -\ 2 \\ \hline \end{array}$	$\begin{array}{r} 1 \\ -\ 1 \\ \hline \end{array}$
$\begin{array}{r} 6 \\ -\ 2 \\ \hline \end{array}$	$\begin{array}{r} 9 \\ -\ 1 \\ \hline \end{array}$	$\begin{array}{r} 7 \\ -\ 0 \\ \hline \end{array}$	$\begin{array}{r} 10 \\ -\ 2 \\ \hline \end{array}$	$\begin{array}{r} 3 \\ -\ 1 \\ \hline \end{array}$

Name ______________________________

Date ______________________________

Guided Class Practice **74A**

***Saxon** Math 2 (for use with **Lesson 74**)*

1. There are 27 children in Room 6, 22 children in Room 7, and 25 children in Room 8. How many children are in the three classes?

Number sentence ______________________________

Answer ______________________________

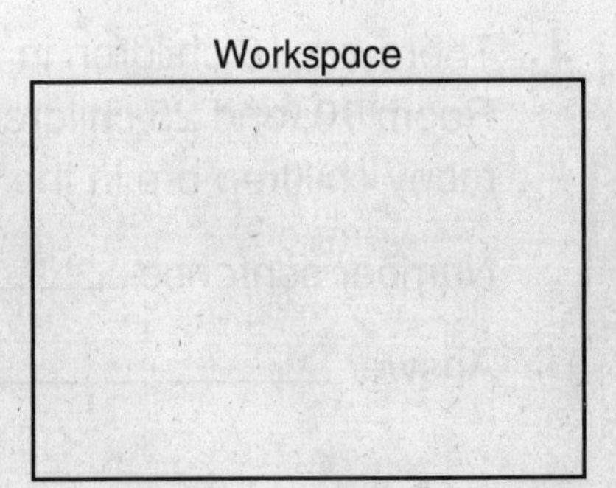

2. Use your ruler to draw these line segments.

$2\frac{1}{2}''$ •

$\frac{1}{2}''$ •

3. Circle the shape that has one half shaded.

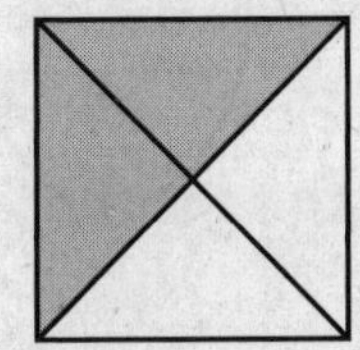

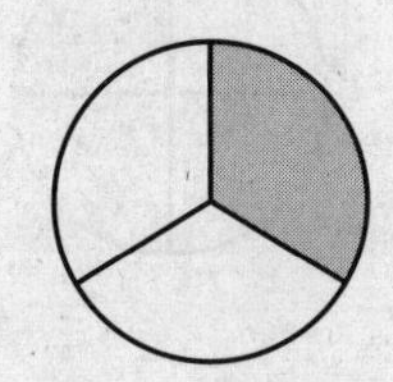

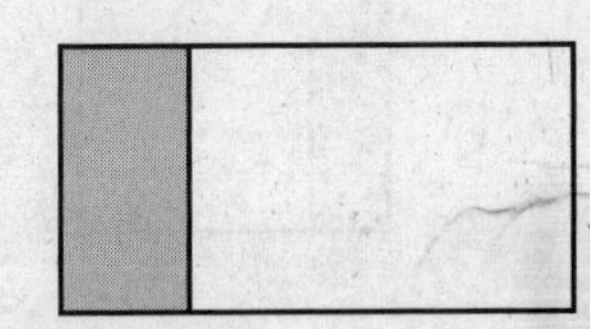

4. One of these is my favorite number.
Cross out the numbers that cannot be my favorite number.

It has 2 digits.
It is between 80 and 100.
It is not an odd number.
What is my favorite number? __________

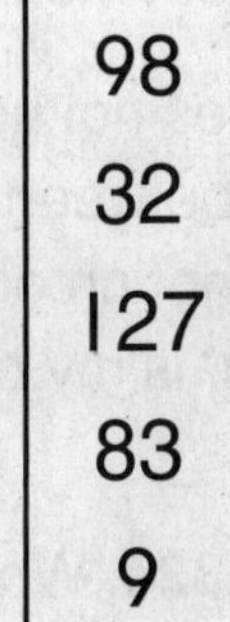

5. I have **38¢.** What coins could I have?

____

What is another way to make **38¢?**

____

6. Fill in the missing numbers on this piece of a hundred number chart.

	34				
				47	

Name ________________________________

Date ________________________________

Homework 74B

***Saxon** Math 2 (for use with **Lesson 74**)*

1. There are 18 children in Room 9, 23 children in Room 10, and 25 children in Room 11. How many children are in the three classes?

Workspace

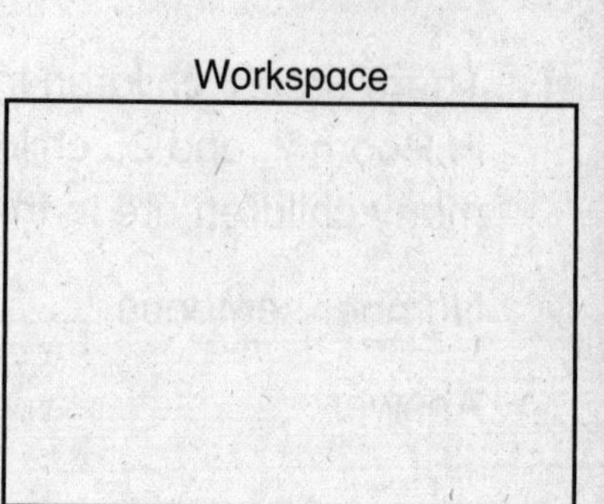

Number sentence ______________________

Answer ______________________

2. Find the answers.

17 − 10 = ______ 35 + 10 = ______ 49 − 10 = ______

13 + 10 = ______ 42 − 10 = ______ 72 + 10 = ______

3. Circle the shape that has one half shaded.

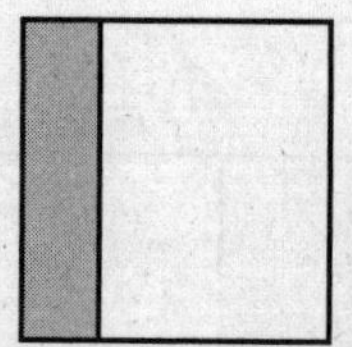

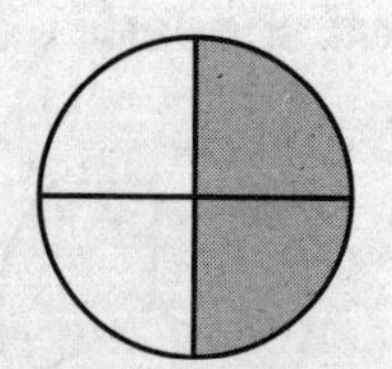

4. One of these is my brother's favorite number.
Cross out the numbers that cannot be his favorite number.

It does not have 3 digits.
It is between 10 and 30.
It is not an odd number.
What is my brother's favorite number? ______

148
24
15
38
82

5. I have **32¢.** What coins could I have?

____

What is another way to make **32¢?**

____

6. Fill in the missing numbers on this piece of a hundred number chart.

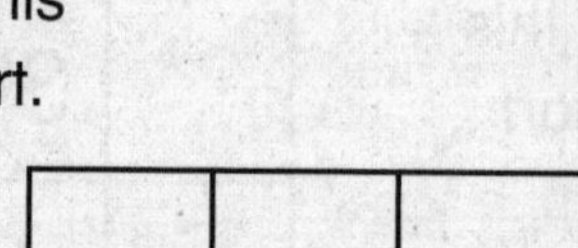

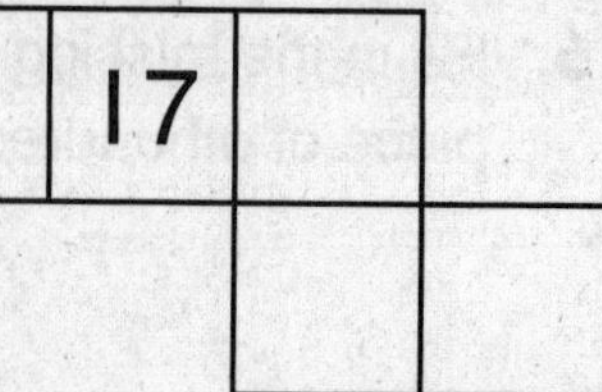

						17		
			25					

Name ______________________________

Fact Assessment 14-1

***Saxon** Math 2 (for use with **Lesson 75-1**)*

Set 14: Subtracting 2, 1, and 0

$\begin{array}{r} 4 \\ -\ 2 \\ \hline \end{array}$	$\begin{array}{r} 1 \\ -\ 1 \\ \hline \end{array}$	$\begin{array}{r} 10 \\ -\ 1 \\ \hline \end{array}$	$\begin{array}{r} 7 \\ -\ 2 \\ \hline \end{array}$	$\begin{array}{r} 3 \\ -\ 1 \\ \hline \end{array}$
$\begin{array}{r} 0 \\ -\ 0 \\ \hline \end{array}$	$\begin{array}{r} 10 \\ -\ 2 \\ \hline \end{array}$	$\begin{array}{r} 6 \\ -\ 1 \\ \hline \end{array}$	$\begin{array}{r} 7 \\ -\ 0 \\ \hline \end{array}$	$\begin{array}{r} 2 \\ -\ 2 \\ \hline \end{array}$
$\begin{array}{r} 9 \\ -\ 2 \\ \hline \end{array}$	$\begin{array}{r} 5 \\ -\ 1 \\ \hline \end{array}$	$\begin{array}{r} 3 \\ -\ 2 \\ \hline \end{array}$	$\begin{array}{r} 8 \\ -\ 1 \\ \hline \end{array}$	$\begin{array}{r} 5 \\ -\ 0 \\ \hline \end{array}$
$\begin{array}{r} 4 \\ -\ 1 \\ \hline \end{array}$	$\begin{array}{r} 11 \\ -\ 2 \\ \hline \end{array}$	$\begin{array}{r} 8 \\ -\ 0 \\ \hline \end{array}$	$\begin{array}{r} 5 \\ -\ 2 \\ \hline \end{array}$	$\begin{array}{r} 9 \\ -\ 1 \\ \hline \end{array}$
$\begin{array}{r} 8 \\ -\ 2 \\ \hline \end{array}$	$\begin{array}{r} 3 \\ -\ 0 \\ \hline \end{array}$	$\begin{array}{r} 7 \\ -\ 1 \\ \hline \end{array}$	$\begin{array}{r} 6 \\ -\ 2 \\ \hline \end{array}$	$\begin{array}{r} 2 \\ -\ 1 \\ \hline \end{array}$

Name ________________________________

Fact Homework 75A

Saxon *Math 2 (for use with* ***Lesson 75-1****)*

Set 14: Subtracting 2, 1, and 0

1. Pretend you are the teacher.
2. Correct this paper.
3. If the answer is incorrect, write the correct answer next to the problem.

$4 - 2 = 2$	$1 - 1 = 0$	$10 - 1 = 9$	$7 - 2 = 5$	$3 - 1 = 4$
$0 - 0 = 0$	$10 - 2 = 7$	$6 - 1 = 5$	$7 - 0 = 7$	$2 - 2 = 0$
$9 - 2 = 7$	$5 - 1 = 4$	$3 - 2 = 1$	$8 - 1 = 6$	$5 - 0 = 5$
$4 - 1 = 3$	$11 - 2 = 8$	$8 - 0 = 8$	$5 - 2 = 3$	$9 - 1 = 8$
$8 - 2 = 6$	$3 - 0 = 3$	$7 - 1 = 6$	$6 - 2 = 8$	$2 - 1 = 1$

Name ______________________________

Fact Homework 75B

***Saxon** Math 2 (for use with **Lesson 75-1**)*

A100: 100 Addition Facts   Corrected by ______

1	$9 + 1$	$2 + 2$	$6 + 4$	$5 + 1$	$0 + 7$	$9 + 9$	$7 + 3$	$1 + 6$	$2 + 5$	$5 + 4$
2	$9 + 4$	$2 + 0$	$8 + 7$	$4 + 1$	$6 + 6$	$7 + 8$	$3 + 2$	$9 + 8$	$0 + 8$	$4 + 6$
3	$5 + 2$	$3 + 9$	$0 + 6$	$8 + 1$	$3 + 3$	$7 + 4$	$7 + 0$	$1 + 5$	$6 + 7$	$2 + 3$
4	$1 + 0$	$5 + 5$	$7 + 6$	$3 + 4$	$2 + 1$	$9 + 5$	$7 + 2$	$4 + 9$	$0 + 3$	$6 + 8$
5	$8 + 2$	$3 + 5$	$1 + 7$	$0 + 0$	$6 + 2$	$5 + 7$	$1 + 4$	$8 + 6$	$2 + 9$	$5 + 0$
6	$6 + 3$	$0 + 5$	$3 + 7$	$4 + 4$	$9 + 2$	$1 + 8$	$6 + 5$	$2 + 4$	$8 + 8$	$0 + 9$
7	$4 + 2$	$7 + 7$	$9 + 0$	$9 + 6$	$5 + 8$	$0 + 1$	$3 + 6$	$7 + 9$	$6 + 0$	$4 + 8$
8	$7 + 1$	$2 + 6$	$4 + 7$	$1 + 2$	$4 + 5$	$8 + 9$	$3 + 0$	$8 + 3$	$1 + 9$	$5 + 6$
9	$1 + 1$	$3 + 8$	$0 + 2$	$5 + 9$	$9 + 3$	$2 + 7$	$8 + 0$	$4 + 3$	$6 + 9$	$1 + 3$
10	$8 + 5$	$4 + 0$	$5 + 3$	$2 + 8$	$3 + 1$	$7 + 5$	$9 + 7$	$0 + 4$	$8 + 4$	$6 + 1$

Name ______________________________

A. Write the answers.

3 – 3 = ______

4 – 3 = ______

5 – 3 = ______

6 – 3 = ______

7 – 3 = ______

8 – 3 = ______

9 – 3 = ______

10 – 3 = ______

11 – 3 = ______

12 – 3 = ______

13 – 3 = ______

14 – 3 = ______

B. Draw lines to connect the problems to the answers.

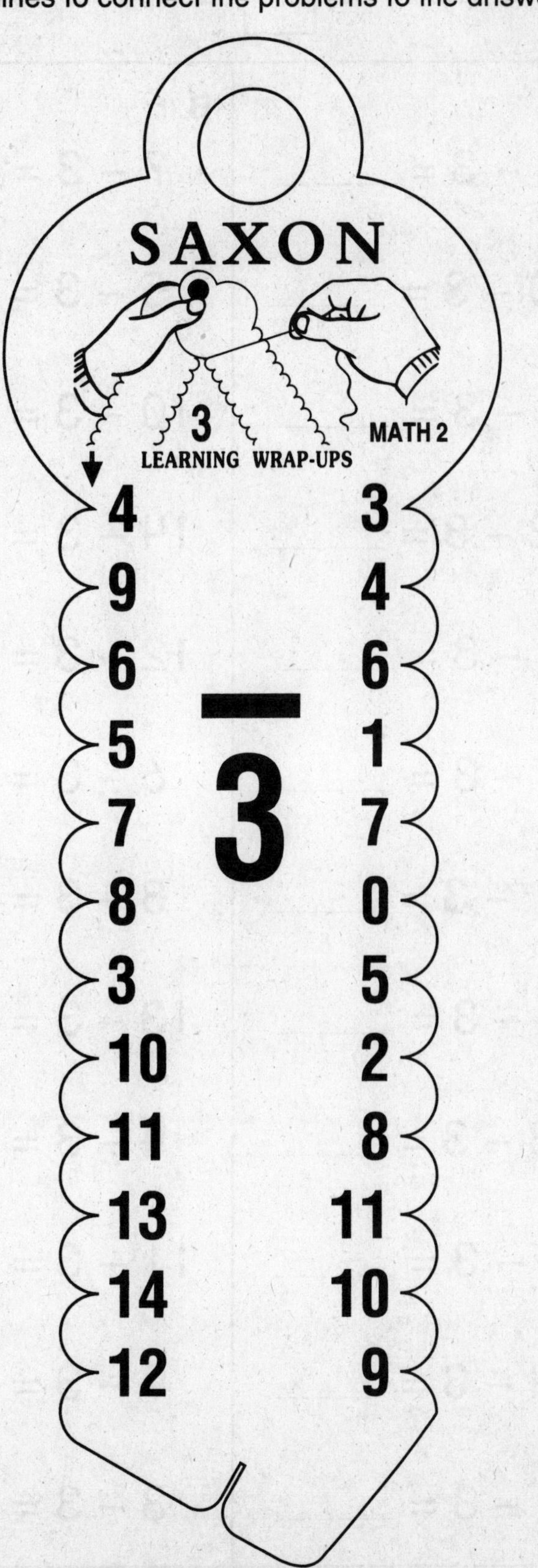

Name ______________________

Do –3 Wrap-Up once. ☐	Do –3 Wrap-Up once. ☐	Do –3 Wrap-Up once. ☐
A.	**B.**	**C.**
6 – 3 = ____	7 – 3 = ____	9 – 3 = ____
10 – 3 = ____	5 – 3 = ____	11 – 3 = ____
3 – 3 = ____	10 – 3 = ____	4 – 3 = ____
12 – 3 = ____	14 – 3 = ____	13 – 3 = ____
9 – 3 = ____	12 – 3 = ____	7 – 3 = ____
7 – 3 = ____	6 – 3 = ____	10 – 3 = ____
11 – 3 = ____	8 – 3 = ____	3 – 3 = ____
4 – 3 = ____	13 – 3 = ____	12 – 3 = ____
13 – 3 = ____	4 – 3 = ____	8 – 3 = ____
8 – 3 = ____	11 – 3 = ____	5 – 3 = ____
14 – 3 = ____	9 – 3 = ____	14 – 3 = ____
5 – 3 = ____	3 – 3 = ____	6 – 3 = ____

Name ______________________

Date ______________________

Guided Class Practice 75A

***Saxon** Math 2 (for use with **Lesson 75-1**)*

1. Erin is on page thirty-five in the book she is reading. If she reads twenty more pages, what page will she be on then? Write a number sentence for the story. Write the answer with a label.

Workspace

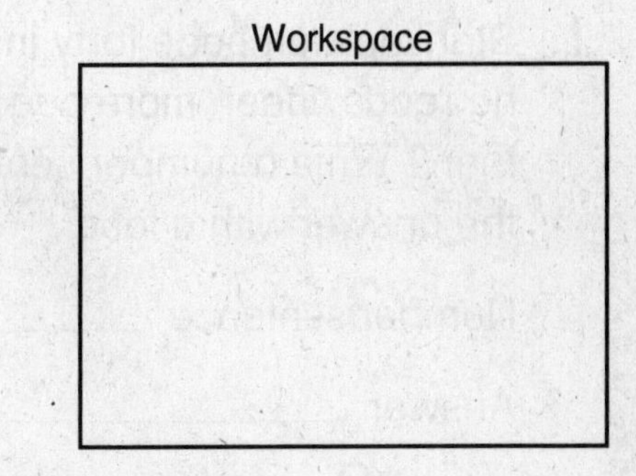

Number sentence ______________________

Answer ______________________

2. Finish numbering the number line using the odd numbers. Put a point at **9.** Label it **M.** Put a point at **15.** Label it **P.**

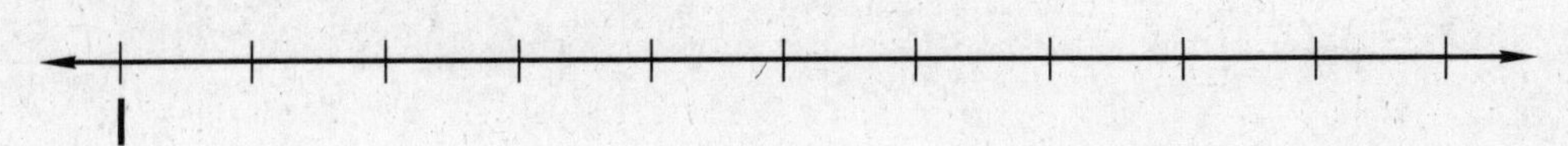

3. I have **42¢.** What coins could I have?

____

What is another way to make **42¢?**

____

4. Color the congruent polygons red.

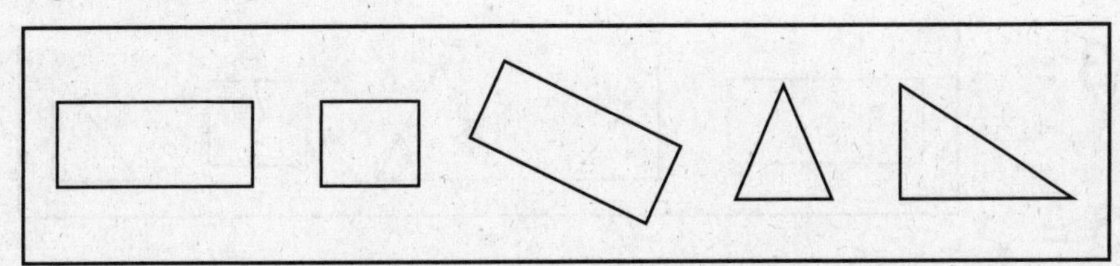

5. About how long is a new pencil?

7 inches　　15 inches　　1 inch　　24 inches

6. Write the problems vertically. Find the sums.

43¢ + 28¢

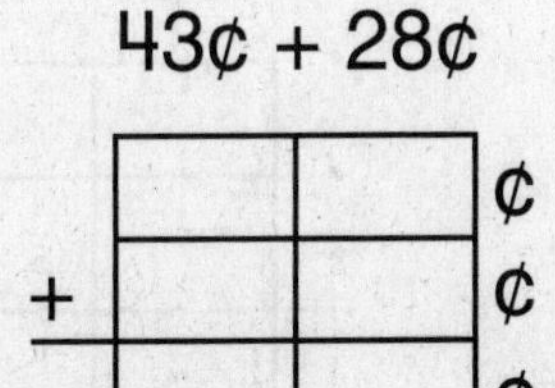

31¢ + 9¢

52¢ + 13¢

Name ______________________

Date ______________________

Homework 75B

Saxon Math 2 (for use with **Lesson 75-1**)

1. Stanley is on page forty in the book he is reading. If he reads fifteen more pages, what page will he be on then? Write a number sentence for the story. Write the answer with a label.

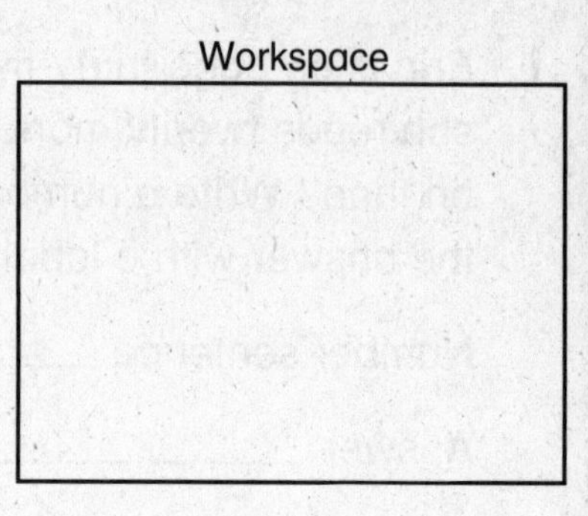

Number sentence ______________________

Answer ______________________

2. Finish numbering the number line using the odd numbers. Put a point at **7.** Label it **T.** Put a point at **11.** Label it **V.**

1

3. I have **45¢.** What coins could I have?

____

What is another way to make **45¢?**

____

4. Color the congruent polygons red.

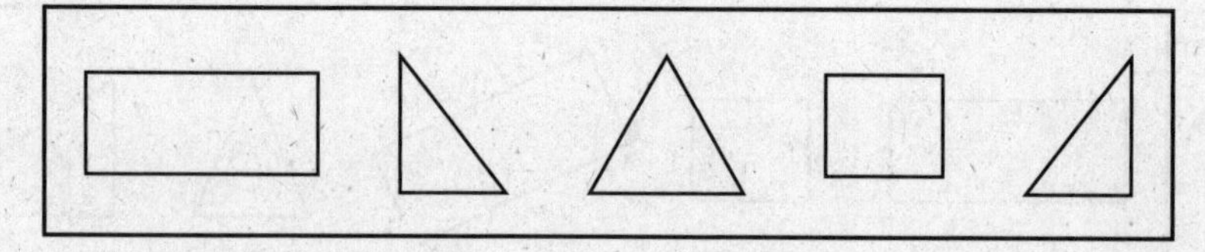

5. About how long is an egg carton?

2 inches   50 inches   30 inches   12 inches

6. Write the problems vertically. Find the sums.

62¢ + 27¢

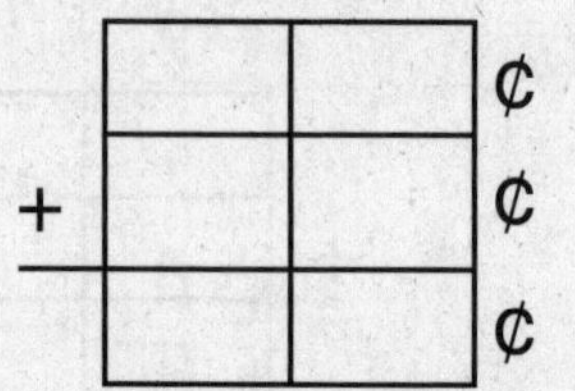

17¢ + 45¢

8¢ + 86¢

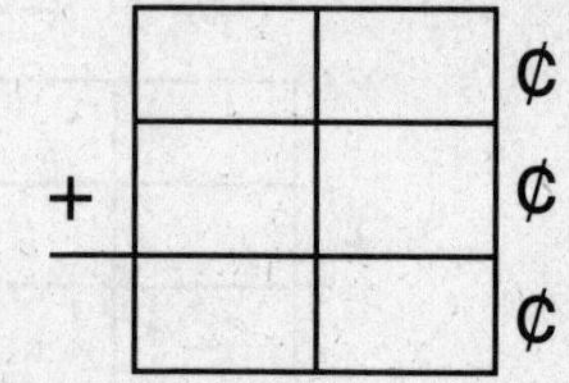

Name ______________________ Score ______

Fact Assessment 14-2

***Saxon** Math 2 (for use with **Lesson 75-2**)*

A100: 100 Addition Facts

1	9 + 1	2 + 2	6 + 4	5 + 1	0 + 7	9 + 9	7 + 3	1 + 6	2 + 5	5 + 4
2	9 + 4	2 + 0	8 + 7	4 + 1	6 + 6	7 + 8	3 + 2	9 + 8	0 + 8	4 + 6
3	5 + 2	3 + 9	0 + 6	8 + 1	3 + 3	7 + 4	7 + 0	1 + 5	6 + 7	2 + 3
4	1 + 0	5 + 5	7 + 6	3 + 4	2 + 1	9 + 5	7 + 2	4 + 9	0 + 3	6 + 8
5	8 + 2	3 + 5	1 + 7	0 + 0	6 + 2	5 + 7	1 + 4	8 + 6	2 + 9	5 + 0
6	6 + 3	0 + 5	3 + 7	4 + 4	9 + 2	1 + 8	6 + 5	2 + 4	8 + 8	0 + 9
7	4 + 2	7 + 7	9 + 0	9 + 6	5 + 8	0 + 1	3 + 6	7 + 9	6 + 0	4 + 8
8	7 + 1	2 + 6	4 + 7	1 + 2	4 + 5	8 + 9	3 + 0	8 + 3	1 + 9	5 + 6
9	1 + 1	3 + 8	0 + 2	5 + 9	9 + 3	2 + 7	8 + 0	4 + 3	6 + 9	1 + 3
10	8 + 5	4 + 0	5 + 3	2 + 8	3 + 1	7 + 5	9 + 7	0 + 4	8 + 4	6 + 1

Name ______________________________

Date ______________________________

1. Ryan has 5 dimes and 7 pennies.
How much money does he have? __________

Daniel has 3 dimes and 2 pennies.
How much money does he have? __________

How much money do the two boys have altogether?

Number sentence ______________________

Answer ______________________

Workspace

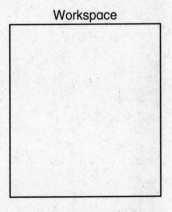

2. Draw a dozen doughnuts. Color a half dozen brown to show that they are chocolate.

How many doughnuts are chocolate? ________

3. Show half past two on the clocks.

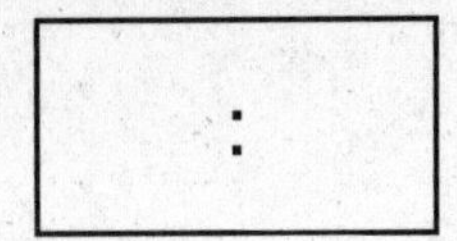

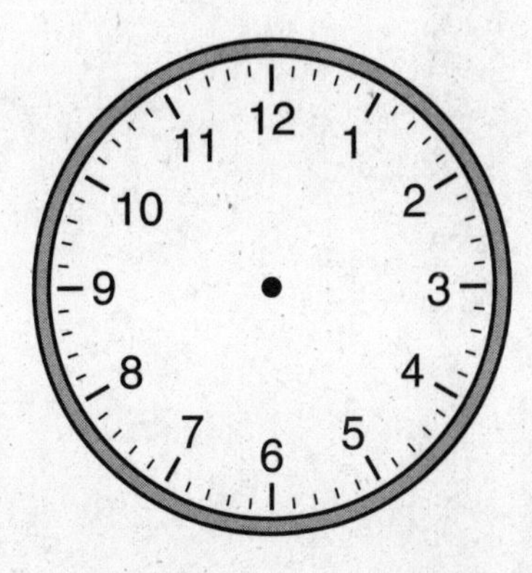

It's morning.
Circle the correct label.

a.m. p.m.

4. I have 5 dimes, 3 nickels, and 2 pennies.
Draw the coins. How much money is this? __________

5. Circle the shape that is congruent to the triangle on the left.

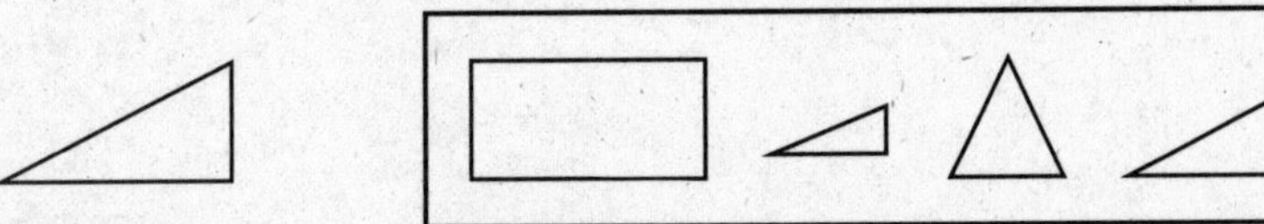

6. Find the answers.

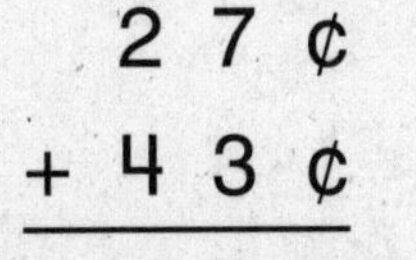

27¢	59¢	13¢	36¢
+ 43¢	+ 26¢	+ 84¢	+ 46¢

Name ____________________

Class Fact Practice **76A**

Saxon *Math 2 (for use with* ***Lesson 76****)*

Set 15: Subtracting 3 and 2

Do –3 Wrap-Up once.
Do –2 Wrap-Up once.

9 – 3	4 – 3	12 – 3	6 – 3	10 – 3
5 – 3	8 – 3	11 – 3	3 – 3	7 – 3
6 – 2	2 – 2	8 – 2	5 – 2	10 – 2
9 – 2	4 – 2	11 – 2	7 – 2	3 – 2
10 – 1	1 – 1	3 – 1	8 – 1	5 – 1

Name ______________________

Set 15: Subtracting 3 and 2

1. Read the answers to someone.
2. Write the answers.
3. Ask someone to correct your paper. Corrected by ______________

$3 - 3$	$12 - 3$	$11 - 3$	$6 - 3$	$8 - 3$
$5 - 3$	$9 - 3$	$7 - 3$	$10 - 3$	$4 - 3$
$11 - 2$	$7 - 2$	$4 - 2$	$8 - 2$	$2 - 2$
$9 - 2$	$6 - 2$	$3 - 2$	$10 - 2$	$5 - 2$
$3 - 1$	$8 - 1$	$9 - 1$	$2 - 1$	$5 - 1$

Name •
Draw a 3-inch line segment.

Date •
Draw a $3\frac{1}{2}$-inch line segment.

Workspace

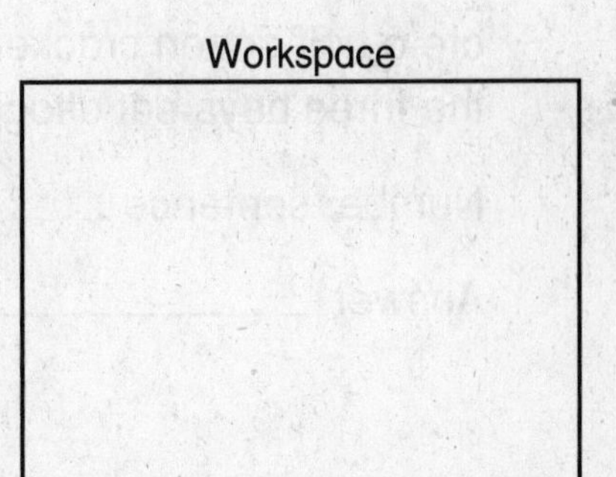

1. Ryan and David each ate a dozen marshmallows. Michael ate a half-dozen marshmallows. How many marshmallows did the three boys eat altogether?

Number sentence ______________________

Answer ______________________

2. Write these numbers in order from least to greatest.

291	67	134	178

_____ least _____ _____ _____ greatest

3. Color the thermometer to show 44°F.

°F
110
100
90
80
70
60
50
40
30
20
10
0
−10
−20

4. Draw a picture to show 251. (Use □ for 100, ▯ for 10, and ▫ for 1.)

5. Use your ruler to measure these line segments.

•————————————————• _____"

•————————• _____"

6. It's morning. What time is it?

Answer ______________________

7. Find the sums.

$$\begin{array}{r} 48¢ \\ +\ 87¢ \\ \hline \end{array} \qquad \begin{array}{r} 53¢ \\ +\ 76¢ \\ \hline \end{array}$$

Name ______________________________

Date ______________________________

Homework 76B

***Saxon** Math 2 (for use with **Lesson 76**)*

1. Joe and Robert each ate a half-dozen crackers. Collin ate a half-dozen crackers. How many crackers did the three boys eat altogether?

Number sentence ______________________

Answer ______________________

Workspace

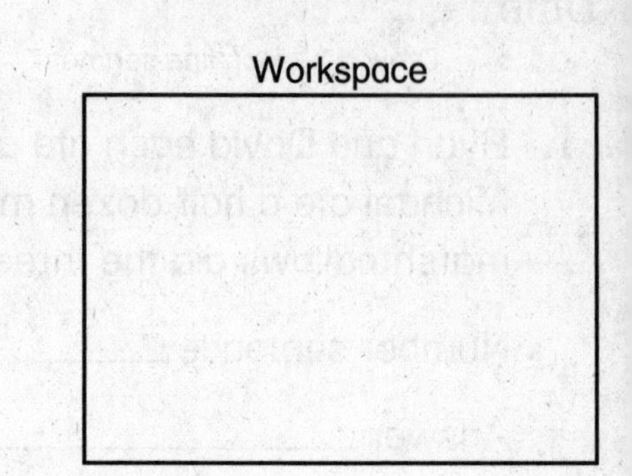

2. Write these numbers in order from least to greatest.

152	212	73	243

_______ least _______ _______ _______ greatest

3. Color the thermometer to show 56°F.

°F
110
100
90
80
70
60
50
40
30
20
10
0
−10
−20

4. Draw a picture to show 143. (Use ☐ for 100, ▯ for 10, and ▪ for 1.)

5. Draw a dozen cookies.
Daniel ate a half dozen.
Put an X on each cookie he ate.

6. It's afternoon. What time is it?

Answer ______________________

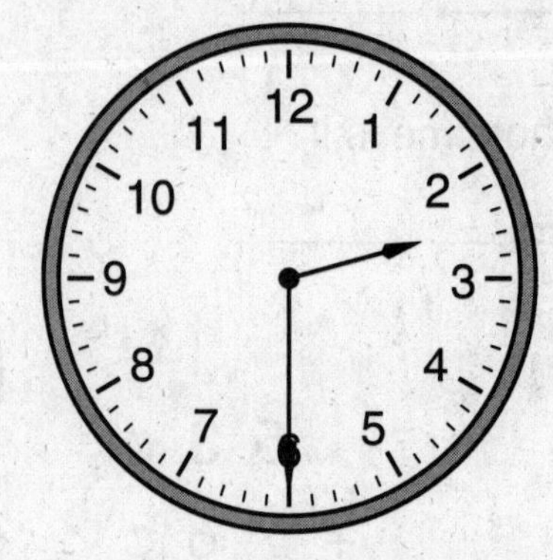

7. Find the sums.

$$\begin{array}{r} 54¢ \\ +\ 95¢ \\ \hline \end{array} \qquad \begin{array}{r} 56¢ \\ +\ 49¢ \\ \hline \end{array}$$

Name ______________________________

Class Fact Practice 77A

Saxon Math 2 (for use with Lesson 77)

Set 15: Subtracting 3 and 2

Do –3 Wrap-Up twice. ☐☐

Do –2 Wrap-Up twice. ☐☐

6 – 3 = ____	2 – 2 = ____
9 – 3 = ____	9 – 2 = ____
12 – 3 = ____	6 – 2 = ____
5 – 3 = ____	10 – 2 = ____
8 – 3 = ____	4 – 2 = ____
10 – 3 = ____	8 – 2 = ____
4 – 3 = ____	11 – 2 = ____
11 – 3 = ____	3 – 2 = ____
7 – 3 = ____	7 – 2 = ____
3 – 3 = ____	5 – 2 = ____

Name ____________________

Saxon Math 2 (for use with **Lesson 77**)

Set 15: Subtracting 3 and 2

1. Write the answers.
2. Draw lines to connect the problems that have the same answer.
3. Ask someone to correct your paper. Corrected by ____________

$9 - 3 =$ ____ •	• $5 - 2 =$ ____
$6 - 3 =$ ____ •	• $3 - 2 =$ ____
$11 - 3 =$ ____ •	• $7 - 2 =$ ____
$4 - 3 =$ ____ •	• $8 - 2 =$ ____
$8 - 3 =$ ____ •	• $4 - 2 =$ ____
$12 - 3 =$ ____ •	• $10 - 2 =$ ____
$5 - 3 =$ ____ •	• $2 - 2 =$ ____
$7 - 3 =$ ____ •	• $11 - 2 =$ ____
$3 - 3 =$ ____ •	• $9 - 2 =$ ____
$10 - 3 =$ ____ •	• $6 - 2 =$ ____

Name ______________________________

Lesson Worksheet 77

***Saxon** Math 2 (for use with **Lesson 77**)*

Write the number shown by each picture.

1.

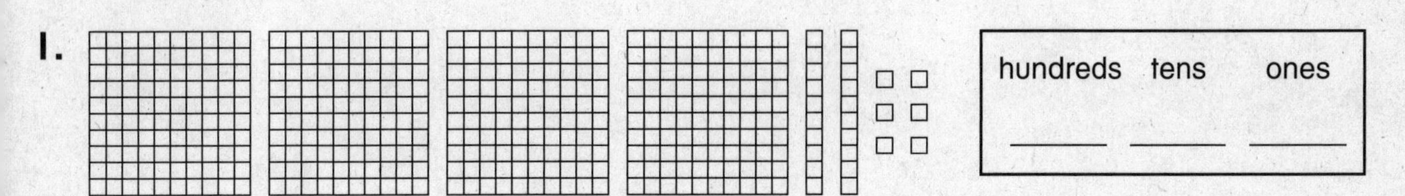

hundreds ______ tens ______ ones ______

2.

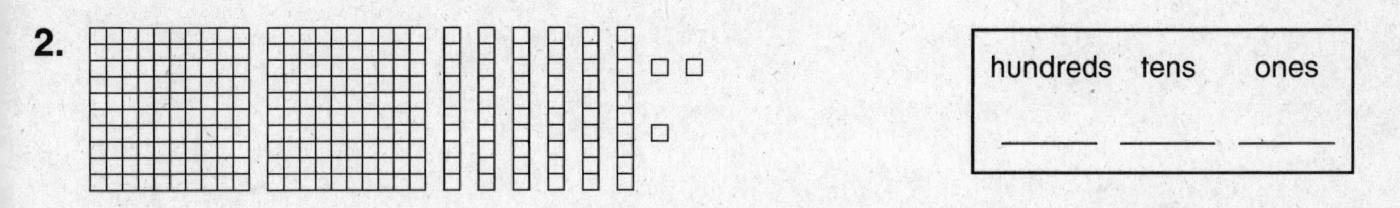

hundreds ______ tens ______ ones ______

3.

hundreds ______ tens ______ ones ______

4.

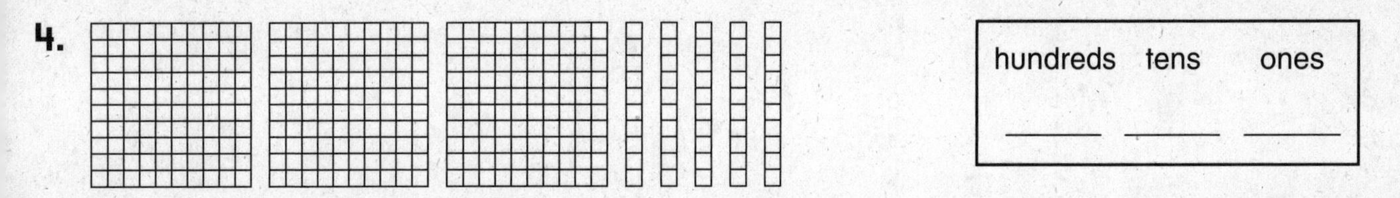

hundreds ______ tens ______ ones ______

5.

____

6.

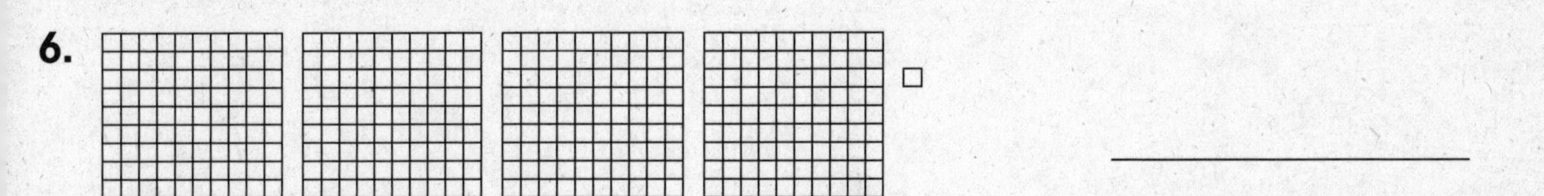

____

7.

____

Name .
Draw a 3-inch line segment.

Date .
Draw a $4\frac{1}{2}$-inch line segment.

Guided Class Practice 77A

***Saxon** Math 2 (for use with **Lesson 77**)*

Workspace

1. On Tuesday fifty-four children in Grade 2 at Washington School ate hot lunch. The other sixty-seven Grade 2 children brought their lunch from home. How many children are in Grade 2 at Washington School?

Number sentence ______________________

Answer ______________________

2. What number does this picture show? ________

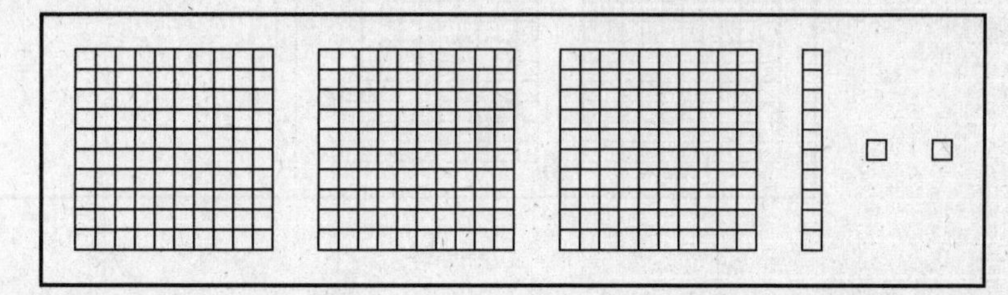

3. Use the graph to answer the questions.

How many children's names are on this graph? ________

How many children like chocolate ice cream? ________

How many children like both vanilla and chocolate ice cream? ________

Which flavor does Carol like? ________

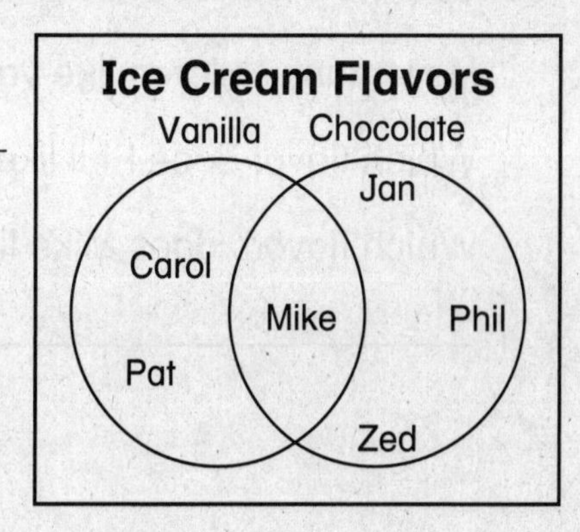

4. Write the fraction that tells how much is shaded. ____

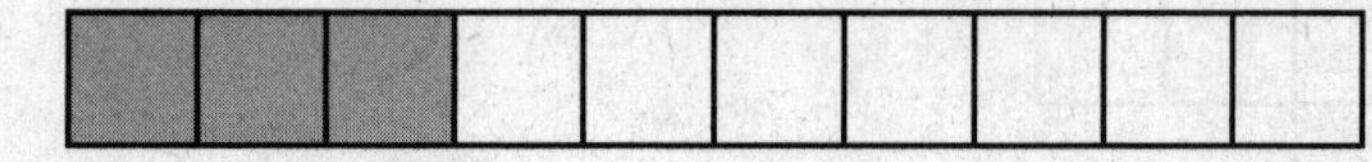

Write the fraction that tells how much is not shaded. ____

5. Find the answers.

10 more than 52 = ________ 10 less than 41 = ________

39 − 10 = ________ 6 + 3 + 2 + 4 + 8 + 1 = ________

$$\begin{array}{r} 47¢ \\ +\ 86¢ \\ \hline \end{array}$$

Name ______________________

Date ______________________

1. On Tuesday sixty-seven children in Grade 2 at Forest School ate hot lunch. The other thirty-five Grade 2 children brought their lunch from home. How many children are in Grade 2 at Forest School?

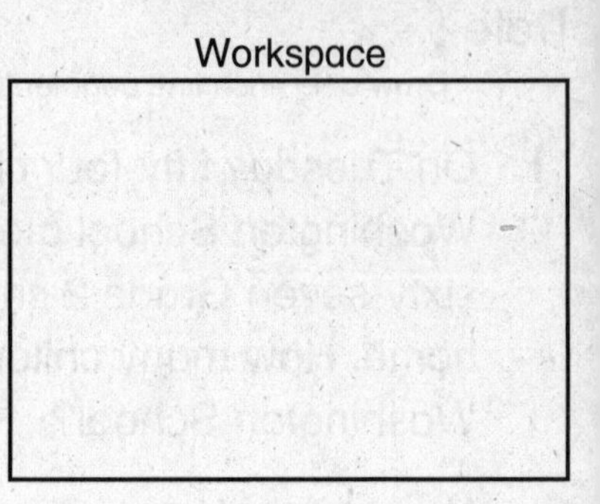

Number sentence ______________________

Answer ______________________

2. What number does this picture show? ________

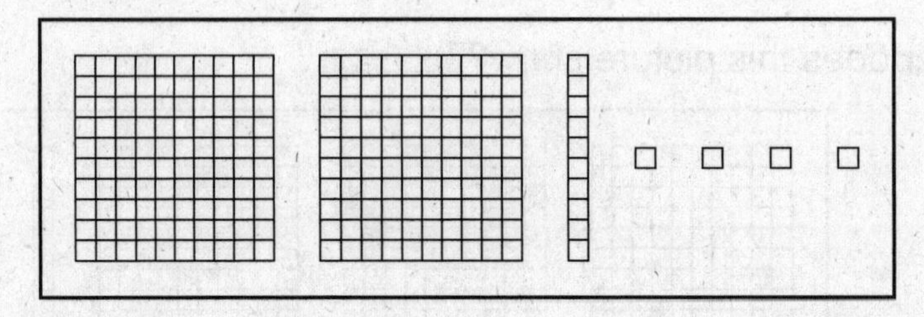

3. Use the graph to answer the questions.

How many children like vanilla ice cream? ________

Which flavor does Phil like? ________________

Which flavors does Mike like? ________________

____

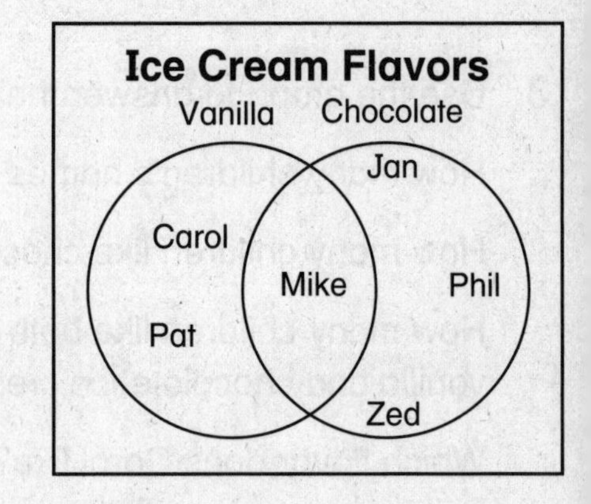

4. Write the fraction that tells how much is shaded. ____

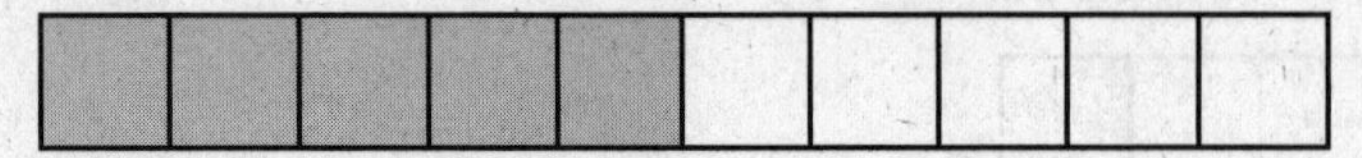

Write the fraction that tells how much is not shaded. ____

5. Find the answers.

10 more than 72 = ________

10 less than 63 = ________

84 − 10 = ________

9 + 3 + 1 + 4 + 7 = ________

  67¢
+ 63¢
______

Name ____________________

Class Fact Practice 78A

***Saxon** Math 2 (for use with **Lesson 78**)*

Set 15: Subtracting 3 and 2

$4 - 3$	$6 - 2$	$9 - 0$	$10 - 3$	$2 - 2$
$10 - 1$	$6 - 3$	$8 - 2$	$12 - 3$	$5 - 2$
$3 - 2$	$8 - 3$	$7 - 2$	$3 - 3$	$9 - 2$
$6 - 1$	$11 - 3$	$4 - 2$	$1 - 0$	$5 - 3$
$9 - 3$	$10 - 2$	$4 - 1$	$7 - 3$	$11 - 2$

Name ____________________

Saxon Math 2 (for use with Lesson 78)

Set 15: Subtracting 3 and 2

1. Read the answers to someone.
2. Write the answers.
3. Ask someone to correct your paper. Corrected by ____________________

$6 - 3$	$7 - 2$	$10 - 3$	$4 - 1$	$9 - 2$
$3 - 2$	$9 - 3$	$6 - 1$	$5 - 2$	$3 - 3$
$10 - 2$	$5 - 3$	$1 - 0$	$4 - 2$	$7 - 3$
$11 - 2$	$4 - 3$	$8 - 2$	$10 - 1$	$11 - 3$
$9 - 0$	$8 - 3$	$6 - 2$	$12 - 3$	$2 - 2$

Name ______________________________

Write the time.

___ : ___

Write the time.

___ : ___

Write the time.

___ : ___

Show the time.

2:45

Show the time.

8:25

Show the time.

1:50

Name •

Draw a 2-inch line segment.

Date •

Draw a $2\frac{1}{2}$-inch line segment.

Guided Class Practice 78A

Saxon Math 2 (for use with Lesson 78)

1. David had a new box of 48 crayons. He gave Bobby 10 crayons to use. How many crayons does David have now?

Number sentence ______________________________

Answer ____________________

2. What is the temperature on the thermometer? ________°F

3. Draw a picture to show 235. (Use ☐ for 100, ▯ for 10, and ▫ for 1.)

4. Fill in the missing numbers on this piece of a hundred number chart.

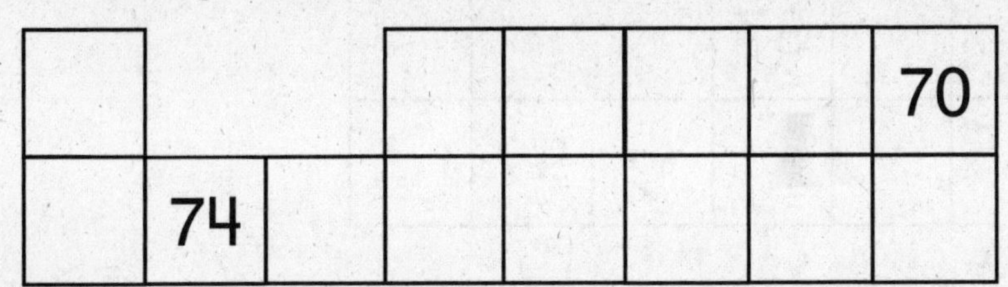

5. It's afternoon.
What time is it? ____________________

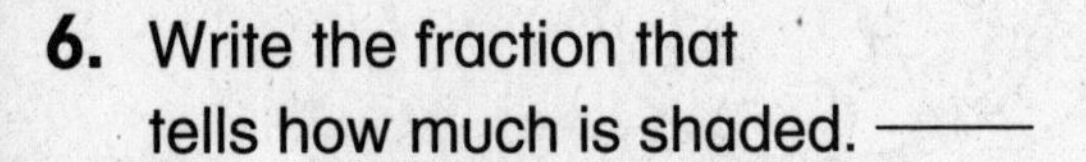

6. Write the fraction that tells how much is shaded. ____

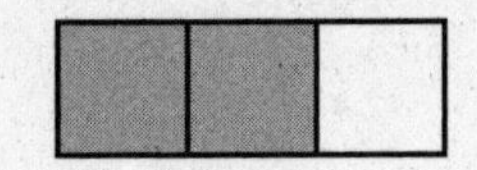

7. Find the sums.

$$\begin{array}{r} 67¢ \\ +\ 92¢ \\ \hline \end{array} \qquad \begin{array}{r} 74¢ \\ +\ 78¢ \\ \hline \end{array} \qquad \begin{array}{r} 39¢ \\ 17¢ \\ +\ 41¢ \\ \hline \end{array}$$

57¢ + 45¢

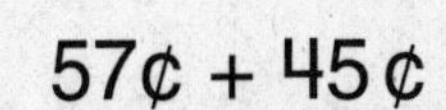

Name ____________________________

Date ____________________________

***Saxon** Math 2 (for use with **Lesson 78**)*

1. Melanie had a new set of 24 markers. She gave Scott 10 markers to use. How many markers does Melanie have now?

Number sentence ____________________________

Answer ____________________

2. What is the temperature on the thermometer? ______°F

°F 110 100 90 80 70 60 50 40 30 20 10 0 −10 −20

3. Draw a picture to show 142. (Use ☐ for 100, ▯ for 10, and ▫ for 1.)

4. Fill in the missing numbers on this piece of a hundred number chart.

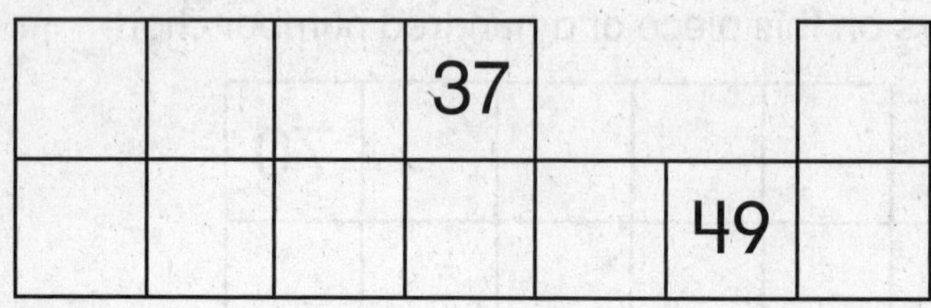

5. It's morning.
What time is it? ____________

6. Write the fraction that tells how much is shaded. ______

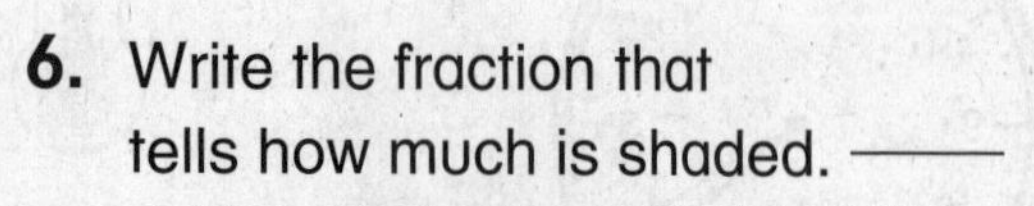

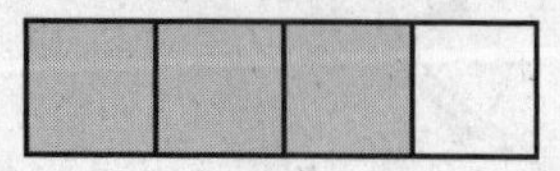

7. Find the sums.

$$\begin{array}{r} 83¢ \\ +\ 65¢ \\ \hline \end{array} \qquad \begin{array}{r} 62¢ \\ +\ 49¢ \\ \hline \end{array} \qquad \begin{array}{r} 35¢ \\ 13¢ \\ +\ 42¢ \\ \hline \end{array}$$

39¢ + 82¢

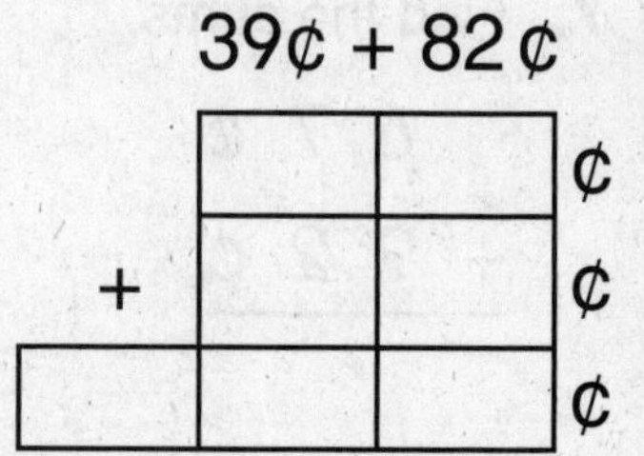

Class Fact Practice 79A

Name ____________________

Saxon Math 2 (for use with Lesson 79)

Set 15: Subtracting 3 and 2

Do −3 Wrap-Up twice.
Do −2 Wrap-Up once.

$5 - 3$	$1 - 0$	$4 - 2$	$11 - 3$	$6 - 1$
$9 - 2$	$3 - 3$	$7 - 2$	$8 - 3$	$3 - 2$
$5 - 2$	$12 - 3$	$8 - 2$	$6 - 3$	$10 - 1$
$2 - 2$	$10 - 3$	$9 - 0$	$6 - 2$	$4 - 3$
$11 - 2$	$7 - 3$	$4 - 1$	$10 - 2$	$9 - 3$

Name ______________________________

Saxon Math 2 (for use with **Lesson 79**)

Set 15: Subtracting 3 and 2

1. Read the answers to someone.
2. Write the answers.
3. Ask someone to correct your paper. Corrected by ______________________

$3 - 3$	$5 - 2$	$6 - 1$	$9 - 3$	$3 - 2$
$9 - 2$	$4 - 1$	$10 - 3$	$7 - 2$	$6 - 3$
$11 - 3$	$10 - 1$	$8 - 2$	$4 - 3$	$11 - 2$
$7 - 3$	$4 - 2$	$1 - 0$	$5 - 3$	$10 - 2$
$2 - 2$	$12 - 3$	$6 - 2$	$8 - 3$	$9 - 0$

Name ________________________________

Add It Up

Round 1

Dollars	Dimes	Pennies
+		

Round 2

Dollars	Dimes	Pennies
+		

Round 3

Dollars	Dimes	Pennies
+		

Amount Won

	Dollars	Dimes	Pennies	
				Round 1
				Round 2
+				Round 3
				Grand Total

Name •
Draw a 3-inch line segment.

Date •
Draw a $3\frac{1}{2}$-inch line segment.

1. There were a dozen children in Room 5. Two children went to the library. How many children are in Room 5 now?

Number sentence ______________________

Answer ______________________

2. Show 5:35 on the clock.
How many minutes are there in one hour? ________

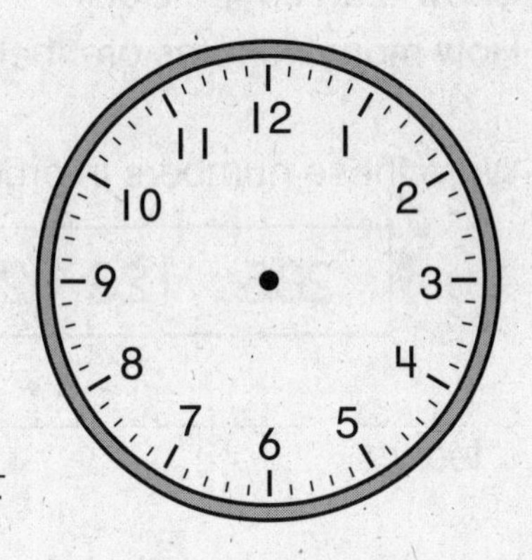

3. Write these numbers in order from least to greatest.

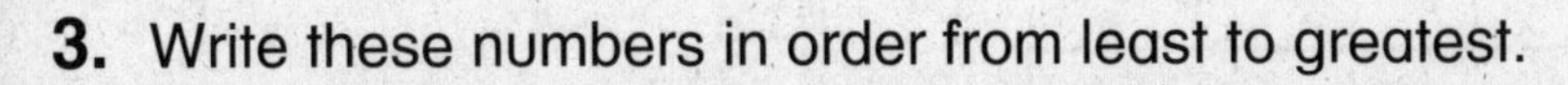

129	243	170	260	259

________ ________ ________ ________ ________
least greatest

4. Write the numbers that are 10 less and 10 more.

_____, 60, _____ _____, 35, _____ _____, 47, _____

5. Using tally marks, show the number of children in your class.

6. What number does this picture show? ________

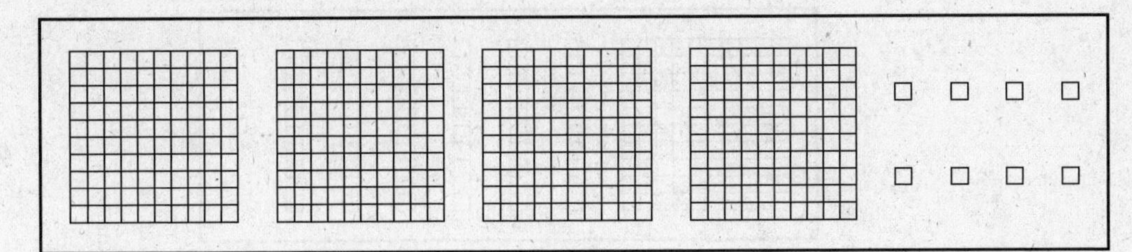

7. Find the answers.

74¢	32¢	55¢
19¢	46¢	28¢
+ 21¢	+ 41¢	+ 85¢

45 − 10 = ________

26 − 10 = ________

M2(3e)-GP-079a

Name ______________________________

Date ______________________________

Saxon Math 2 (for use with Lesson 79)

1. There were 15 children in a swimming pool. Six children climbed out of the pool. How many children are in the pool now?

   Number sentence ______________________________

   Answer ______________________________

2. Show 8:25 on the clock.
   How many minutes are there in one hour? ________

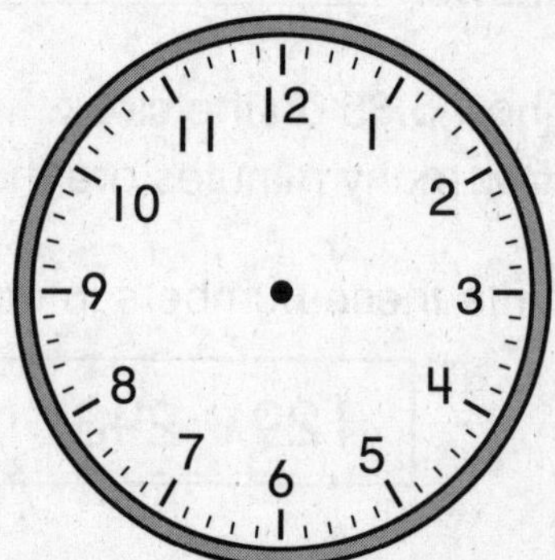

3. Write these numbers in order from least to greatest.

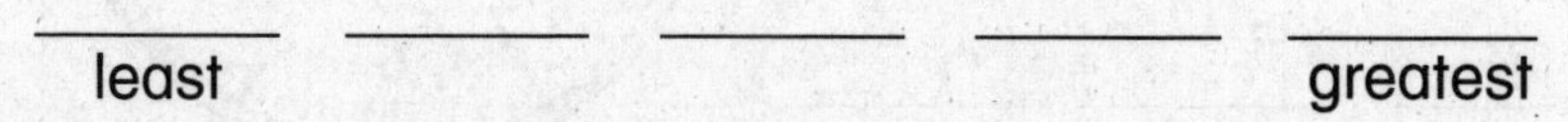

________ ________ ________ ________ ________
least   greatest

4. Write the numbers that are 10 less and 10 more.

   ______, 80, ______   ______, 55, ______   ______, 39, ______

5. Using tally marks, show the number of lights in your house.

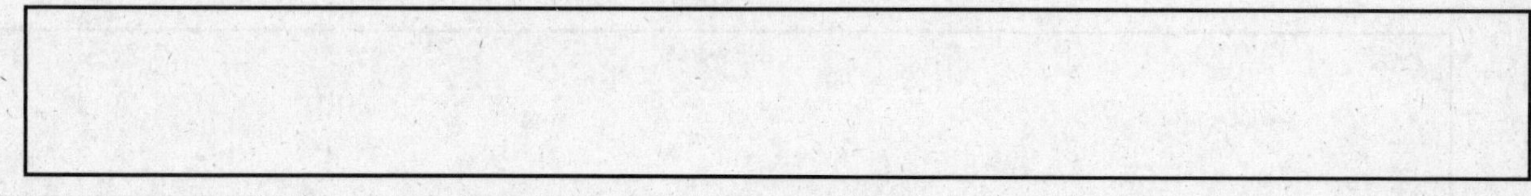

6. What number does this picture show? ________

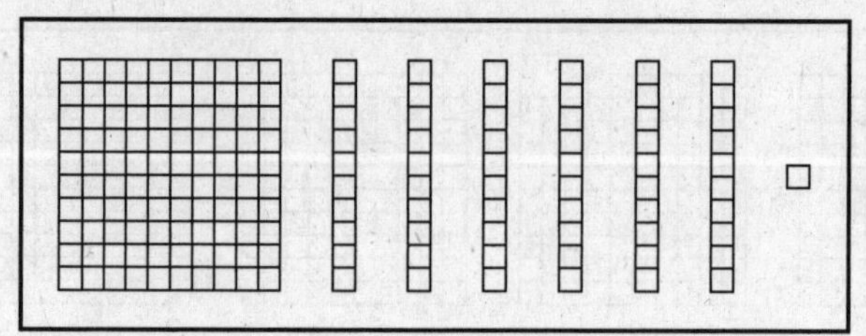

7. Find the answers.

28¢	31¢	36¢
57¢	94¢	64¢
+ 42¢	+ 13¢	+ 33¢

84 − 10 = ________

75 − 10 = ________

Name ______________________ Score ______

Fact Assessment 15-1

***Saxon** Math 2 (for use with **Lesson 80-1**)*

Set 15: Subtracting 3 and 2

$\begin{array}{r} 6 \\ -\ 3 \\ \hline \end{array}$	$\begin{array}{r} 8 \\ -\ 2 \\ \hline \end{array}$	$\begin{array}{r} 10 \\ -\ 3 \\ \hline \end{array}$	$\begin{array}{r} 1 \\ -\ 0 \\ \hline \end{array}$	$\begin{array}{r} 3 \\ -\ 3 \\ \hline \end{array}$
$\begin{array}{r} 5 \\ -\ 2 \\ \hline \end{array}$	$\begin{array}{r} 9 \\ -\ 3 \\ \hline \end{array}$	$\begin{array}{r} 6 \\ -\ 1 \\ \hline \end{array}$	$\begin{array}{r} 4 \\ -\ 3 \\ \hline \end{array}$	$\begin{array}{r} 10 \\ -\ 2 \\ \hline \end{array}$
$\begin{array}{r} 3 \\ -\ 2 \\ \hline \end{array}$	$\begin{array}{r} 4 \\ -\ 1 \\ \hline \end{array}$	$\begin{array}{r} 8 \\ -\ 3 \\ \hline \end{array}$	$\begin{array}{r} 6 \\ -\ 2 \\ \hline \end{array}$	$\begin{array}{r} 12 \\ -\ 3 \\ \hline \end{array}$
$\begin{array}{r} 7 \\ -\ 2 \\ \hline \end{array}$	$\begin{array}{r} 5 \\ -\ 3 \\ \hline \end{array}$	$\begin{array}{r} 9 \\ -\ 2 \\ \hline \end{array}$	$\begin{array}{r} 2 \\ -\ 2 \\ \hline \end{array}$	$\begin{array}{r} 10 \\ -\ 1 \\ \hline \end{array}$
$\begin{array}{r} 11 \\ -\ 3 \\ \hline \end{array}$	$\begin{array}{r} 4 \\ -\ 2 \\ \hline \end{array}$	$\begin{array}{r} 9 \\ -\ 0 \\ \hline \end{array}$	$\begin{array}{r} 7 \\ -\ 3 \\ \hline \end{array}$	$\begin{array}{r} 11 \\ -\ 2 \\ \hline \end{array}$

Name ______________________________

Fact Homework 80A

***Saxon** Math 2 (for use with **Lesson 80-1**)*

Set 15: Subtracting 3 and 2

Pretend you are the teacher.
Correct this paper.
If the answer is incorrect, write the correct answer next to the problem.

$\begin{array}{r} 6 \\ -\ 3 \\ \hline 3 \end{array}$	$\begin{array}{r} 8 \\ -\ 2 \\ \hline 6 \end{array}$	$\begin{array}{r} 10 \\ -\ 3 \\ \hline 7 \end{array}$	$\begin{array}{r} 1 \\ -\ 0 \\ \hline 1 \end{array}$	$\begin{array}{r} 3 \\ -\ 3 \\ \hline 0 \end{array}$
$\begin{array}{r} 5 \\ -\ 2 \\ \hline 3 \end{array}$	$\begin{array}{r} 9 \\ -\ 3 \\ \hline 5 \end{array}$	$\begin{array}{r} 6 \\ -\ 1 \\ \hline 5 \end{array}$	$\begin{array}{r} 4 \\ -\ 3 \\ \hline 1 \end{array}$	$\begin{array}{r} 10 \\ -\ 2 \\ \hline 8 \end{array}$
$\begin{array}{r} 3 \\ -\ 2 \\ \hline 1 \end{array}$	$\begin{array}{r} 4 \\ -\ 1 \\ \hline 3 \end{array}$	$\begin{array}{r} 8 \\ -\ 3 \\ \hline 6 \end{array}$	$\begin{array}{r} 6 \\ -\ 2 \\ \hline 4 \end{array}$	$\begin{array}{r} 12 \\ -\ 3 \\ \hline 8 \end{array}$
$\begin{array}{r} 7 \\ -\ 2 \\ \hline 5 \end{array}$	$\begin{array}{r} 5 \\ -\ 3 \\ \hline 2 \end{array}$	$\begin{array}{r} 9 \\ -\ 2 \\ \hline 7 \end{array}$	$\begin{array}{r} 2 \\ -\ 2 \\ \hline 4 \end{array}$	$\begin{array}{r} 10 \\ -\ 1 \\ \hline 9 \end{array}$
$\begin{array}{r} 11 \\ -\ 3 \\ \hline 9 \end{array}$	$\begin{array}{r} 4 \\ -\ 2 \\ \hline 2 \end{array}$	$\begin{array}{r} 9 \\ -\ 0 \\ \hline 9 \end{array}$	$\begin{array}{r} 7 \\ -\ 3 \\ \hline 4 \end{array}$	$\begin{array}{r} 11 \\ -\ 2 \\ \hline 9 \end{array}$

Name ______________________

Fact Homework 80B

***Saxon** Math 2 (for use with **Lesson 80-1**)*

A100: 100 Addition Facts Corrected by ______

1	9 + 1	2 + 2	6 + 4	5 + 1	0 + 7	9 + 9	7 + 3	1 + 6	2 + 5	5 + 4
2	9 + 4	2 + 0	8 + 7	4 + 1	6 + 6	7 + 8	3 + 2	9 + 8	0 + 8	4 + 6
3	5 + 2	3 + 9	0 + 6	8 + 1	3 + 3	7 + 4	7 + 0	1 + 5	6 + 7	2 + 3
4	1 + 0	5 + 5	7 + 6	3 + 4	2 + 1	9 + 5	7 + 2	4 + 9	0 + 3	6 + 8
5	8 + 2	3 + 5	1 + 7	0 + 0	6 + 2	5 + 7	1 + 4	8 + 6	2 + 9	5 + 0
6	6 + 3	0 + 5	3 + 7	4 + 4	9 + 2	1 + 8	6 + 5	2 + 4	8 + 8	0 + 9
7	4 + 2	7 + 7	9 + 0	9 + 6	5 + 8	0 + 1	3 + 6	7 + 9	6 + 0	4 + 8
8	7 + 1	2 + 6	4 + 7	1 + 2	4 + 5	8 + 9	3 + 0	8 + 3	1 + 9	5 + 6
9	1 + 1	3 + 8	0 + 2	5 + 9	9 + 3	2 + 7	8 + 0	4 + 3	6 + 9	1 + 3
10	8 + 5	4 + 0	5 + 3	2 + 8	3 + 1	7 + 5	9 + 7	0 + 4	8 + 4	6 + 1

Name ______________________________

Lesson Worksheet 80-1A

***Saxon** Math 2 (for use with **Lesson 80-1**)*

A. Write the answers.

4 – 4 = ______

5 – 4 = ______

6 – 4 = ______

7 – 4 = ______

8 – 4 = ______

9 – 4 = ______

10 – 4 = ______

11 – 4 = ______

12 – 4 = ______

13 – 4 = ______

14 – 4 = ______

15 – 4 = ______

B. Draw lines to connect the problems to the answers.

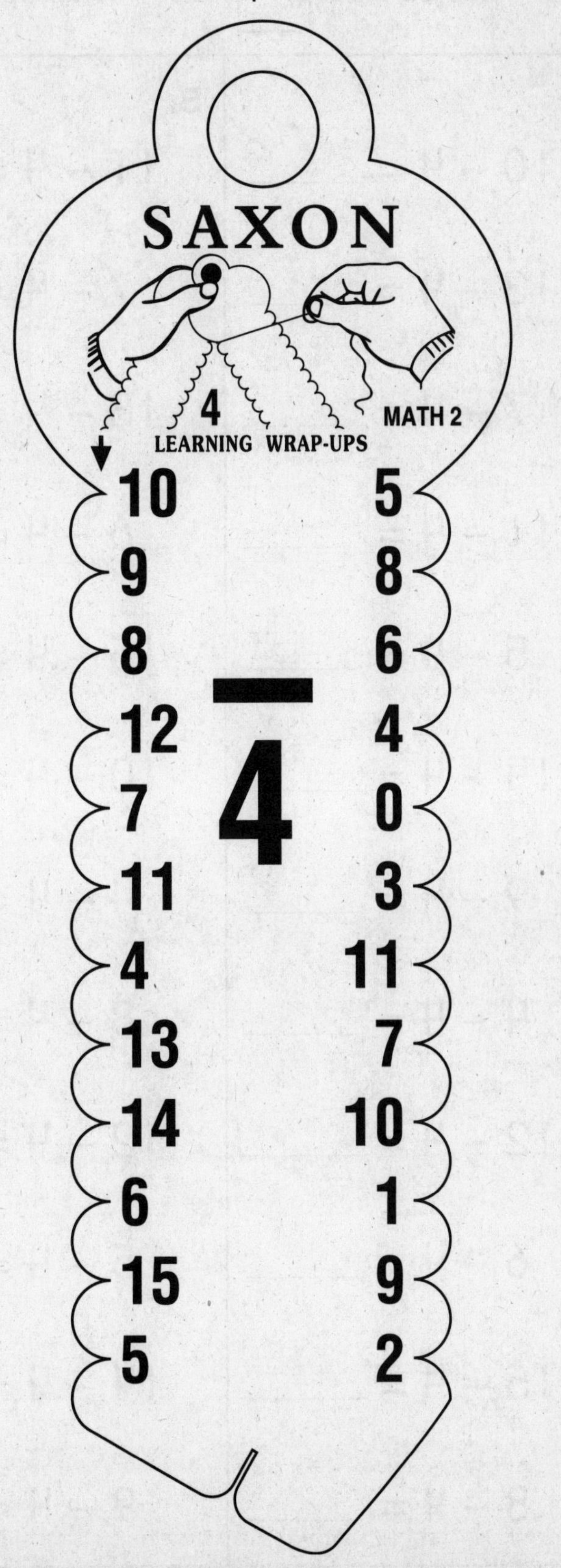

Name ______________________

Saxon Math 2 (for use with **Lesson 80-1**)

Do –4 Wrap-Up once. ☐ Do –4 Wrap-Up once. ☐ Do –4 Wrap-Up once. ☐

A.	**B.**	**C.**
10 – 4 = ____	11 – 4 = ____	5 – 4 = ____
13 – 4 = ____	7 – 4 = ____	8 – 4 = ____
7 – 4 = ____	13 – 4 = ____	12 – 4 = ____
11 – 4 = ____	6 – 4 = ____	9 – 4 = ____
5 – 4 = ____	15 – 4 = ____	4 – 4 = ____
14 – 4 = ____	10 – 4 = ____	13 – 4 = ____
9 – 4 = ____	4 – 4 = ____	7 – 4 = ____
4 – 4 = ____	8 – 4 = ____	10 – 4 = ____
12 – 4 = ____	12 – 4 = ____	6 – 4 = ____
6 – 4 = ____	5 – 4 = ____	14 – 4 = ____
15 – 4 = ____	14 – 4 = ____	11 – 4 = ____
8 – 4 = ____	9 – 4 = ____	15 – 4 = ____

Name ____________________

Date ____________________

Problem-Solving Worksheet 80A

***Saxon** Math 2 (for use with **Lesson 80-1**)*

Understand	Plan	Solve	Check

Draw a Picture

There are square cafeteria tables at Crowley Elementary School. Four children can sit at each table. Show how many tables Mrs. Kavanaugh will need for the 23 children in her class.

How many tables will Mrs. Kavanaugh need for her class? ______

Name ______________________

Understand	Plan	Solve	Check

There are round cafeteria tables at George Clark Elementary School. Five children can sit at each table. Show how many tables Mrs. Tollett will need for the 22 children in her class.

How many tables will Mrs. Tollett need for her class? ________

Circle the problem-solving strategies you used to solve this problem.

Act It Out ***Use Logical Reasoning***

Draw a Picture ***Look for a Pattern***

Make an Organized List ***Guess and Check***

Explain how you got your answer: ______________________

Name ____________________ Score ______

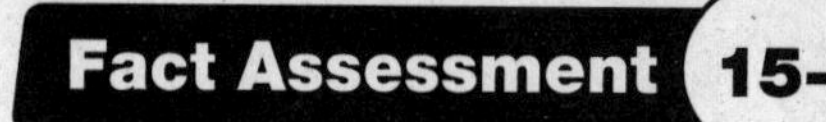

***Saxon** Math 2 (for use with **Lesson 80-2**)*

A100: 100 Addition Facts

1	9 + 1	2 + 2	6 + 4	5 + 1	0 + 7	9 + 9	7 + 3	1 + 6	2 + 5	5 + 4
2	9 + 4	2 + 0	8 + 7	4 + 1	6 + 6	7 + 8	3 + 2	9 + 8	0 + 8	4 + 6
3	5 + 2	3 + 9	0 + 6	8 + 1	3 + 3	7 + 4	7 + 0	1 + 5	6 + 7	2 + 3
4	1 + 0	5 + 5	7 + 6	3 + 4	2 + 1	9 + 5	7 + 2	4 + 9	0 + 3	6 + 8
5	8 + 2	3 + 5	1 + 7	0 + 0	6 + 2	5 + 7	1 + 4	8 + 6	2 + 9	5 + 0
6	6 + 3	0 + 5	3 + 7	4 + 4	9 + 2	1 + 8	6 + 5	2 + 4	8 + 8	0 + 9
7	4 + 2	7 + 7	9 + 0	9 + 6	5 + 8	0 + 1	3 + 6	7 + 9	6 + 0	4 + 8
8	7 + 1	2 + 6	4 + 7	1 + 2	4 + 5	8 + 9	3 + 0	8 + 3	1 + 9	5 + 6
9	1 + 1	3 + 8	0 + 2	5 + 9	9 + 3	2 + 7	8 + 0	4 + 3	6 + 9	1 + 3
10	8 + 5	4 + 0	5 + 3	2 + 8	3 + 1	7 + 5	9 + 7	0 + 4	8 + 4	6 + 1

Name ______________________

Date ______________________

Written Assessment 15

***Saxon** Math 2 (for use with **Lesson 80-2**)*

1. There were a dozen cupcakes in the box. Sara ate one cupcake. How many cupcakes are left?

Number sentence ______________________

Answer ______________________

2. Use the graph to answer the questions.

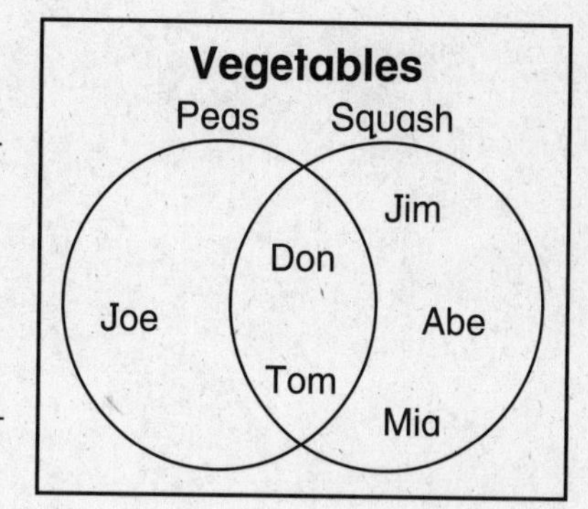

How many children's names are on this graph? ______

How many children like peas? ______

How many children like squash? ______

How many children like both peas and squash? ______

Which vegetable does Jim like? ______

3. What temperature does the thermometer show? ______°F

°F
110
100
90
80
70
60
50
40
30
20
10
0
−10
−20

4. Write the fractions that tell how much is shaded.

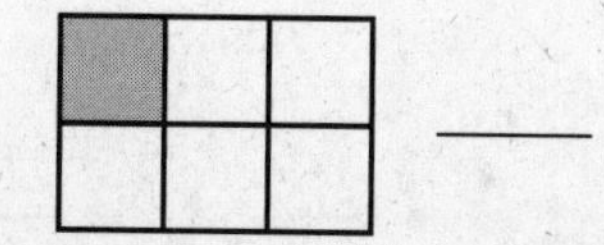 ____

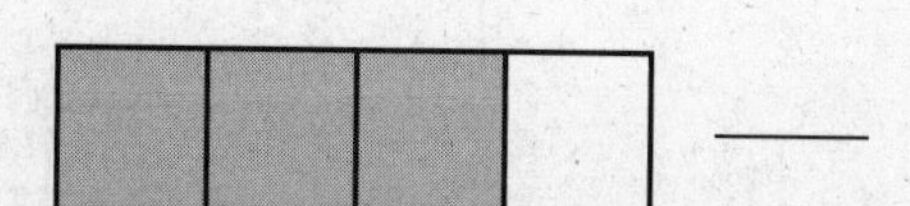 ____

5. Draw a $4\frac{1}{2}$" line segment.

•

Draw a $2\frac{1}{2}$" line segment.

•

6. Find the answers.

56¢	68¢	24¢
+ 32¢	+ 17¢	+ 36¢

42¢ + 19¢

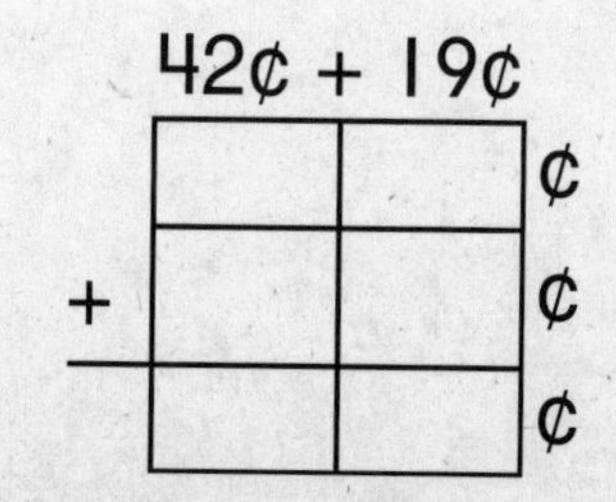

Name ______________________________

Lesson Worksheet 80-2A

Saxon *Math 2 (for use with* ***Lesson 80-2****)*

Use 2 large triangles to cover each polygon.

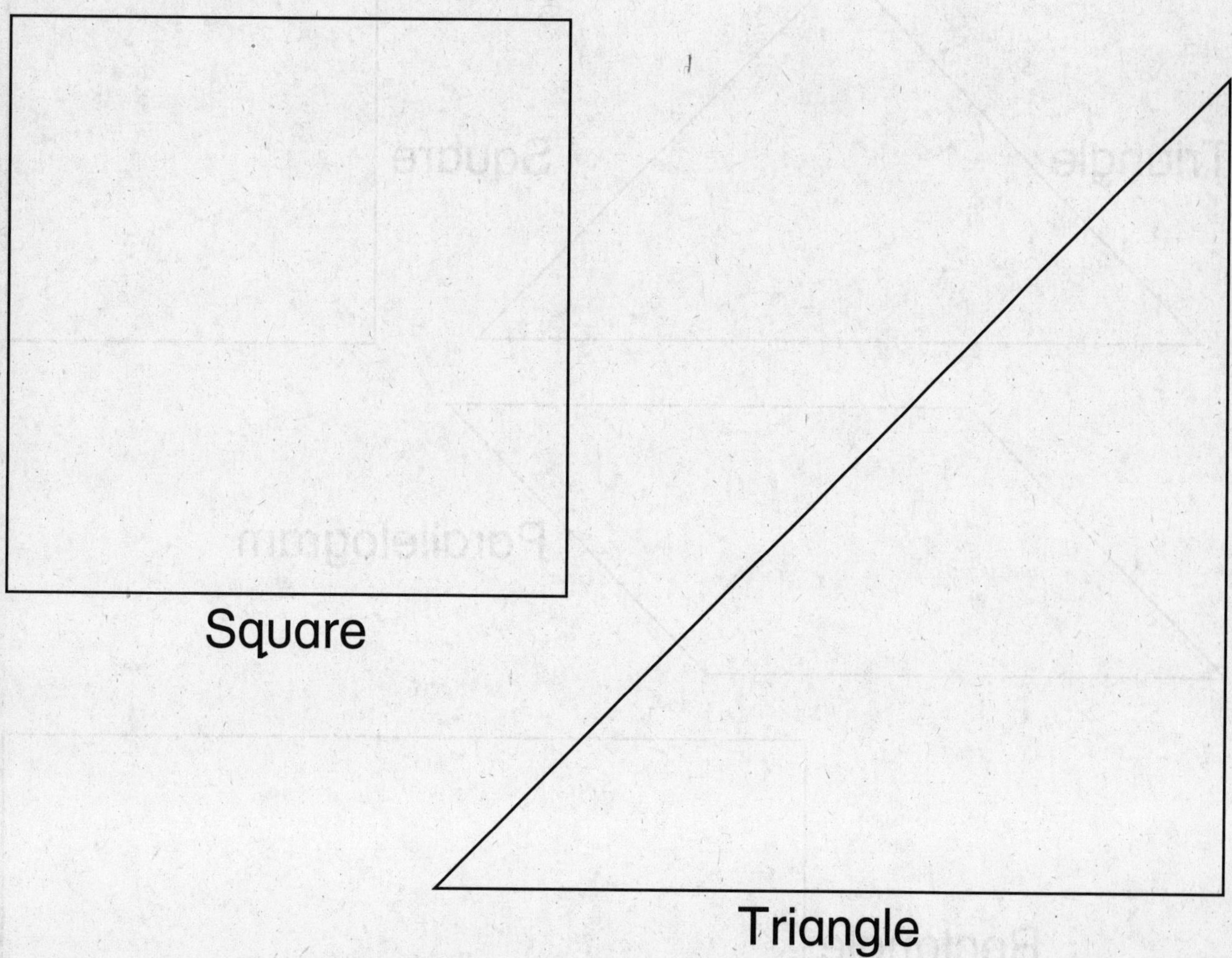

Parallelogram

Saxon Math 2 (for use with Lesson 80-2)

Triangle

Square

Parallelogram

Rectangle

Trapezoid

Name ______________________________

Class Fact Practice 81A

*Saxon Math 2 (for use with **Lesson 81**)*

Set 16: Subtracting 4 and 3; Review Facts

Do –4 Wrap-Up once.
Do –3 Wrap-Up once.

$10 - 4$ $4 - 4$ $12 - 4$ $8 - 4$ $6 - 4$

$7 - 4$ $13 - 4$ $5 - 4$ $11 - 4$ $9 - 4$

$9 - 3$ $5 - 3$ $11 - 3$ $7 - 3$ $3 - 3$

$8 - 3$ $12 - 3$ $4 - 3$ $10 - 3$ $6 - 3$

$10 - 2$ $4 - 2$ $5 - 2$ $8 - 2$ $2 - 2$

Name ______________________________

Saxon Math 2 (for use with Lesson 81)

Set 16: Subtracting 4 and 3; Review Facts

1. Read the answers to someone.
2. Write the answers.
3. Ask someone to correct your paper. Corrected by ______________________

4 − 4	8 − 4	13 − 4	6 − 4	10 − 4
12 − 4	9 − 4	5 − 4	11 − 4	7 − 4
7 − 3	3 − 3	12 − 3	5 − 3	9 − 3
4 − 3	11 − 3	8 − 3	10 − 3	6 − 3
7 − 2	3 − 2	4 − 2	9 − 2	2 − 2

Name

Draw a $2\frac{1}{2}$" line segment.

Date

Measure this line segment using inches. ____"

Guided Class Practice 81A

Saxon Math 2 (for use with **Lesson 81**)

1. Blake bought a marker for 48¢ and an eraser for 37¢. How much money did he spend?

Number sentence ____________________

Answer ____________________

Workspace

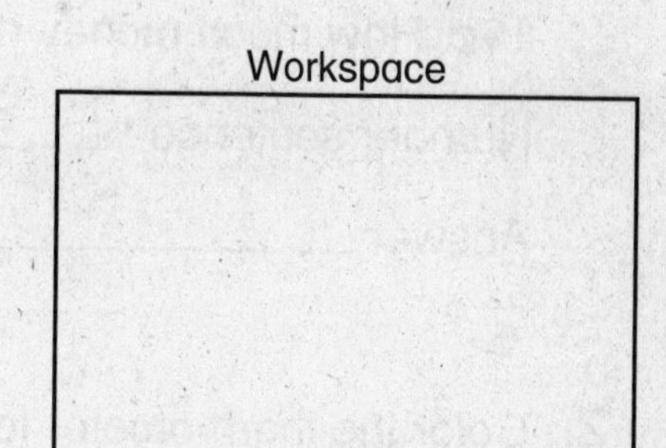

2. Color the thermometer to show 54°F.

°F

110, 100, 90, 80, 70, 60, 50, 40, 30, 20, 10, 0, −10, −20

3. Write these numbers in order from least to greatest.

265 319 256

______ least ______ ______ greatest

4. Draw a picture to show 304. (Use □ for 100, ▯ for 10, and ▫ for 1.)

5. Fill in the correct comparison symbol (>, <, or =).

7 ◯ 9 17 ◯ 6 100 ◯ 200

6. Circle the number that is closest to 80.

54 67 83 95

7. Find the sums.

$$\begin{array}{r} 64 \\ 28 \\ +\ 46 \\ \hline \end{array} \qquad \begin{array}{r} 13 \\ 97 \\ +\ 52 \\ \hline \end{array} \qquad 35¢ + 56¢$$

Name ______________________

Date ______________________

Saxon Math 2 (for use with Lesson 81)

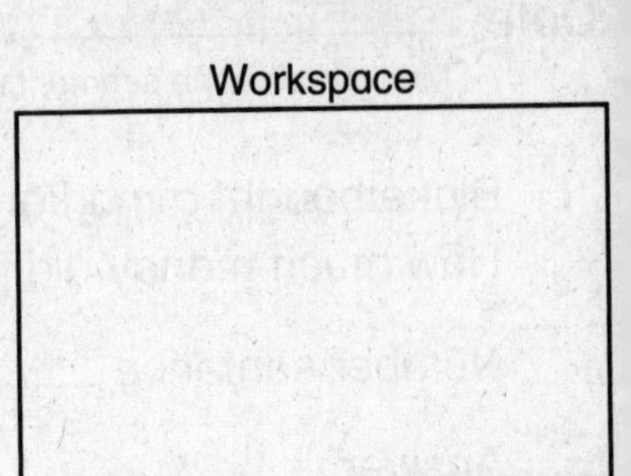

1. Alexis bought a note pad for 64¢ and a pencil for 19¢. How much money did she spend?

   Number sentence ______________________

   Answer ______________________

2. Color the thermometer to show 38°F.

3. Write these numbers in order from least to greatest.

   452 573 537

   ______ ______ ______
   least greatest

4. Draw a picture to show 350. (Use ☐ for 100, ▯ for 10, and ▫ for 1.)

5. Fill in the correct comparison symbol (>, <, or =).

   12 ◯ 7 15 ◯ 18 230 ◯ 179

6. Circle the number that is closest to 70.

   49 55 72 75

7. Find the sums.

   23 + 54 + 47 = ____

   36 + 45 + 35 = ____

   59¢ + 33¢

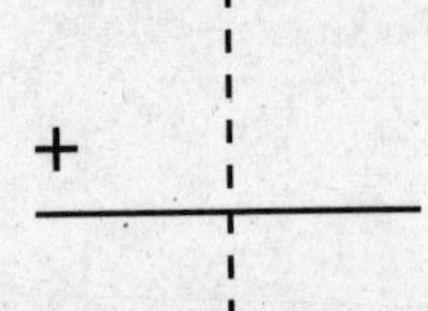

Name ____________________

Class Fact Practice 82A

***Saxon** Math 2 (for use with **Lesson 82**)*

Set 16: Subtracting 4 and 3

Do –4 Wrap-Up twice.

Do –3 Wrap-Up twice.

7 – 4 = ____	3 – 3 = ____
11 – 4 = ____	10 – 3 = ____
13 – 4 = ____	6 – 3 = ____
4 – 4 = ____	12 – 3 = ____
10 – 4 = ____	4 – 3 = ____
8 – 4 = ____	9 – 3 = ____
5 – 4 = ____	7 – 3 = ____
12 – 4 = ____	11 – 3 = ____
9 – 4 = ____	5 – 3 = ____
6 – 4 = ____	8 – 3 = ____

Name ______________________________

Set 16: Subtracting 4 and 3

1. Write the answers.
2. Draw lines to connect the problems that have the same answer.
3. Ask someone to correct your paper. Corrected by ________________

7 − 4 = _____ •	• 4 − 3 = _____
4 − 4 = _____ •	• 8 − 3 = _____
12 − 4 = _____ •	• 9 − 3 = _____
9 − 4 = _____ •	• 6 − 3 = _____
5 − 4 = _____ •	• 11 − 3 = _____
10 − 4 = _____ •	• 3 − 3 = _____
6 − 4 = _____ •	• 7 − 3 = _____
13 − 4 = _____ •	• 10 − 3 = _____
8 − 4 = _____ •	• 5 − 3 = _____
11 − 4 = _____ •	• 12 − 3 = _____

Name ______________________________

Lesson Worksheet 82

***Saxon** Math 2 (for use with **Lesson 82**)*

Doughnuts Eaten

Room Number	Doughnuts
1	
2	
3	8

Room 1	◎ ◎ ◎ ◎ ◎ ◎ ◎
Room 2	◎ ◎ ◎ ◎ (
Room 3	

◎ = 2 doughnuts

Cartons of Milk Drunk

Room Number	Cartons of Milk
1	
2	
3	7

Room 1	
Room 2	
Room 3	

= 2 cartons of milk

Number of Pockets in Our Classroom

Group	Number of Pockets
A	
B	
C	
D	

Group A	
Group B	
Group C	
Group D	

= 2 pockets

Name _______________ Measure this line segment using inches. ______"

Guided Class Practice 82A

***Saxon** Math 2 (for use with **Lesson 82**)*

Date
Draw a $2\frac{1}{2}$" line segment.

Workspace

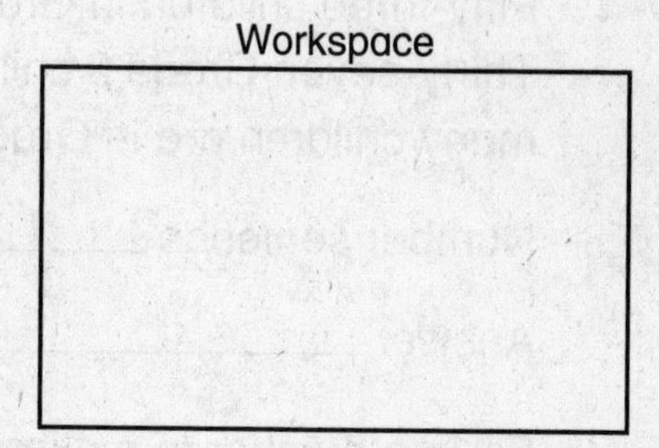

1. Forty-seven children voted no and thirty-nine children voted yes. How many children voted?

Number sentence ______________________

Answer ______________________

2. Fill in a number to make each number sentence true.

$\square > 7$ $\quad$ $5 > \square$ $\quad$ $6 + 4 = \square + 6$

3. How many children chose fall as their favorite season? ______

How many children chose spring? ______

How many children chose summer? ______

Draw pictures to show that 4 children chose winter.

Favorite Seasons

Spring	☺ ☺ ☺
Summer	☺ ☺ ☺ ☺ ◖
Fall	☺
Winter	

☺ = 2 children

4. Find the answers.

$42 - 10 =$ ______ $\quad$ ten more than 26 = ______

$58 + 10 =$ ______ $\quad$ ten less than 47 = ______

$6 + 6 + 4 + 2 =$ ______ $\quad$ 1 less than 12 = ______

5. What numbers are shown by these pictures?

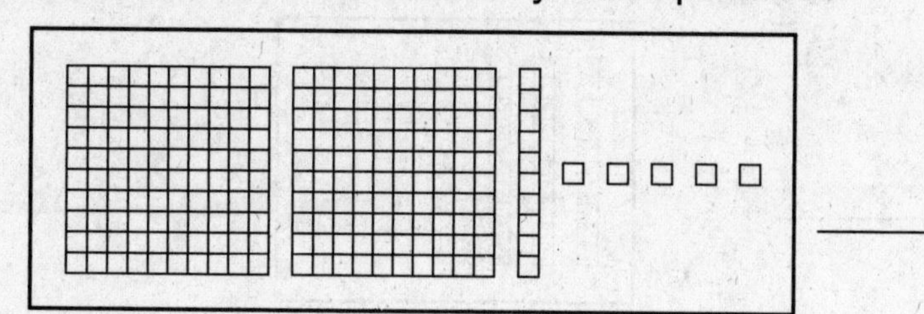
______

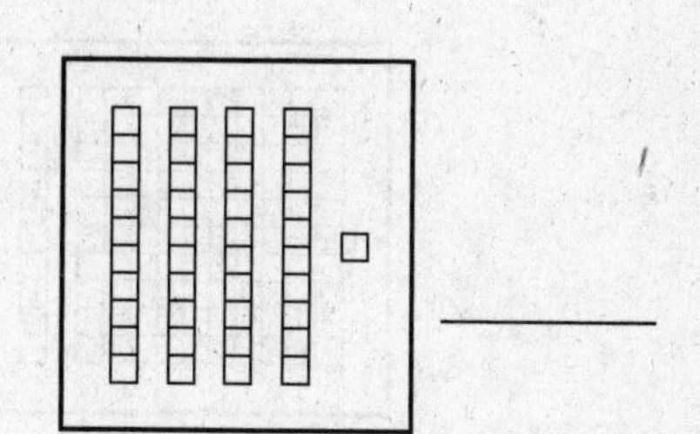
______

6. Find the sums.

$$\begin{array}{r} 59 \\ +\ 73 \\ \hline \end{array} \qquad \begin{array}{r} 72 \\ +\ 85 \\ \hline \end{array} \qquad \begin{array}{r} 42 \\ 87 \\ +\ 21 \\ \hline \end{array} \qquad \begin{array}{r} 35 \\ 54 \\ +\ 69 \\ \hline \end{array}$$

Name ______________________________

Date ______________________________

***Saxon** Math 2 (for use with **Lesson 82**)*

Workspace

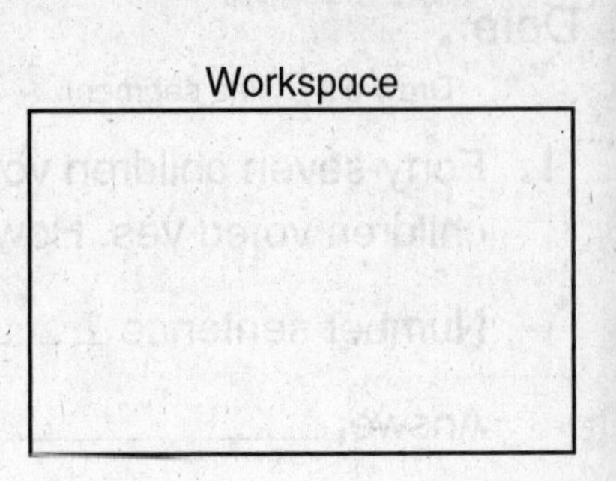

1. Fifty-three children in Grade 2 walk to school. Thirty-seven Grade 2 children ride to school. How many children are in Grade 2?

Number sentence ______________________

Answer ______________________

2. Fill in a number to make each number sentence true.

$4 >$ 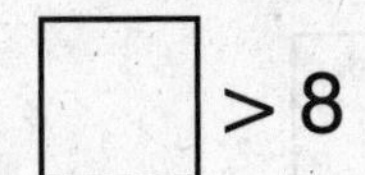 $\square > 8$ $8 + 9 = 9 + \square$

3. How many children chose fall as their favorite season? ______

How many children chose spring? ______

How many children chose summer? ______

Draw pictures to show that 6 children chose winter.

Favorite Seasons

4. Find the answers.

$38 - 10 =$ ______

$59 + 10 =$ ______

$7 + 7 + 3 + 1 =$ ______

ten more than 86 = ______

ten less than 53 = ______

1 less than 14 = ______

5. What numbers are shown by these pictures?

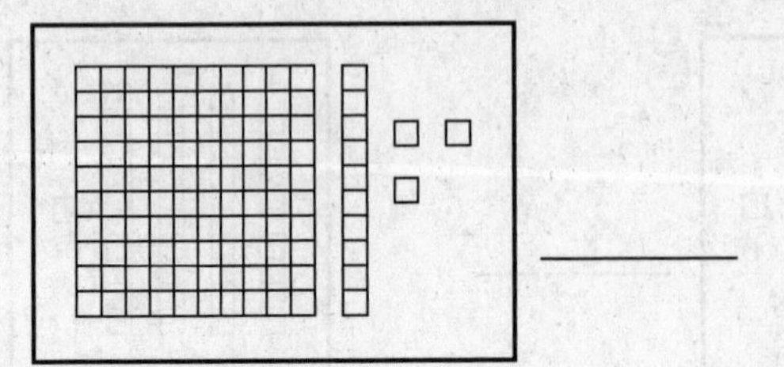 ______

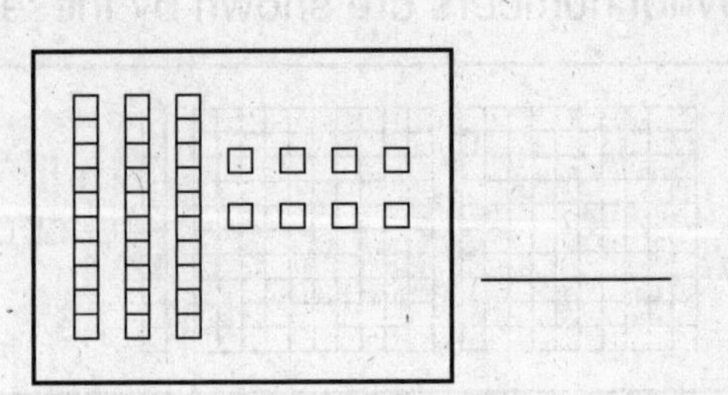 ______

6. Find the sums.

$$\begin{array}{r} 46 \\ +\ 92 \\ \hline \end{array} \qquad \begin{array}{r} 65 \\ +\ 28 \\ \hline \end{array} \qquad \begin{array}{r} 53 \\ 20 \\ +\ 38 \\ \hline \end{array} \qquad \begin{array}{r} 79 \\ 27 \\ +\ 53 \\ \hline \end{array}$$

Name ______________________

Class Fact Practice 83A

***Saxon** Math 2 (for use with **Lesson 83**)*

Set 16: Subtracting 4 and 3

$10 - 4$	$8 - 3$	$4 - 4$	$13 - 4$	$5 - 3$
$11 - 3$	$12 - 4$	$7 - 3$	$11 - 4$	$7 - 4$
$10 - 3$	$5 - 4$	$6 - 3$	$13 - 4$	$10 - 4$
$9 - 4$	$4 - 3$	$12 - 3$	$6 - 4$	$9 - 3$
$12 - 4$	$8 - 4$	$3 - 3$	$11 - 4$	$9 - 4$

Name ________________________________

Saxon Math 2 (for use with Lesson 83)

Set 16: Subtracting 4 and 3

1. Read the answers to someone.
2. Write the answers.
3. Ask someone to correct your paper. Corrected by ______________________

$9 - 4$	$4 - 3$	$11 - 4$	$7 - 4$	$13 - 4$
$12 - 3$	$4 - 4$	$5 - 3$	$10 - 4$	$8 - 3$
$12 - 4$	$11 - 3$	$8 - 4$	$10 - 3$	$9 - 4$
$6 - 3$	$11 - 4$	$9 - 3$	$5 - 4$	$13 - 4$
$10 - 4$	$6 - 4$	$3 - 3$	$12 - 4$	$7 - 3$

Name ______________________

Lesson Worksheet 83A

***Saxon** Math 2 (for use with **Lesson 83**)*

R = Red Y = Yellow B = Blue

1.

R Y

What fractional part is each color?

Red ______ Yellow ______

2.

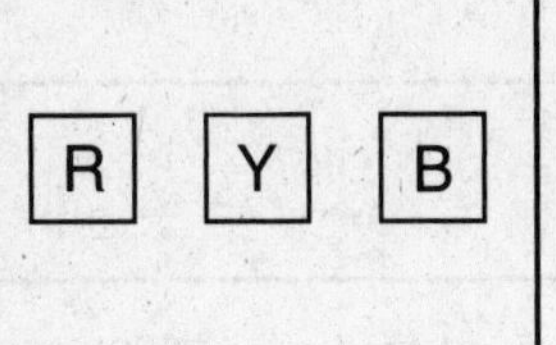

What fractional part is each color?

Red ______ Yellow ______ Blue ______

3.

R Y
B Y

What fractional part is each color?

Red ______ Yellow ______ Blue ______

4.

R R B
Y Y

What fractional part is each color?

Red ______ Yellow ______ Blue ______

5.

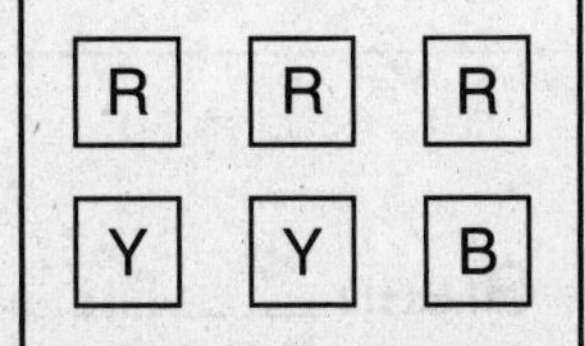

What fractional part is each color?

Red ______ Yellow ______ Blue ______

Name ____________________

6. Draw 2 hearts.

Make $\frac{1}{2}$ of the hearts red.

Make $\frac{1}{2}$ of the hearts blue.

7. Draw 3 stars.

Make $\frac{1}{3}$ of the stars yellow.

Make $\frac{2}{3}$ of the stars blue.

8. Draw 4 flowers.

Make $\frac{1}{4}$ of the flowers red.

Make $\frac{1}{4}$ of the flowers blue.

Make $\frac{2}{4}$ of the flowers yellow.

9. Draw 5 markers.

Make $\frac{3}{5}$ of the markers blue.

Make $\frac{2}{5}$ of the markers yellow.

10. Draw your 12 color tiles.

What fractional part is each color?

Blue ______ Yellow ______ Green ______

Red ______ Brown ______ Orange ______

Name.
Draw a 3-inch line segment.

Date.
Draw a $3\frac{1}{2}$" line segment.

1. There were 25 children in Room 12. Ten children left to go to the library. How many children are in Room 12 now?

Number sentence ____________________

Answer ____________________

2. What fractional part of the circles is shaded? ______

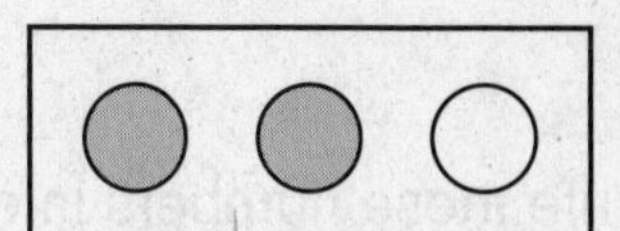

3. Write these numbers in order from least to greatest.

49	434	193	216	180

______ ______ ______ ______ ______
least greatest

4. Use the pictograph to answer these questions.

How many children chose red? ______

How many children chose blue? ______

How many more children chose blue than red? ______

Draw pictures to show that 5 children chose green.

Favorite Colors

Red	☺ ☺ ☺
Blue	☺ ☺ ☺ ☺
Green	

☺ = 2 children

5. It's morning.
What time is it?

____

It's evening.
What time is it?

____

6. Find the sums.

$$\begin{array}{r} 26 \\ +\ 46 \\ \hline \end{array} \qquad \begin{array}{r} 87 \\ +\ 42 \\ \hline \end{array} \qquad \begin{array}{r} 54 \\ 32 \\ +\ 24 \\ \hline \end{array} \qquad \begin{array}{r} 16 \\ 43 \\ +\ 28 \\ \hline \end{array}$$

Name ______________________________

Date ______________________________

1. Sarah had 65¢. She gave her brother a dime. How much money does she have now?

   Number sentence ______________________________

   Answer ______________________________

2. What fractional part of the circles is shaded? ______

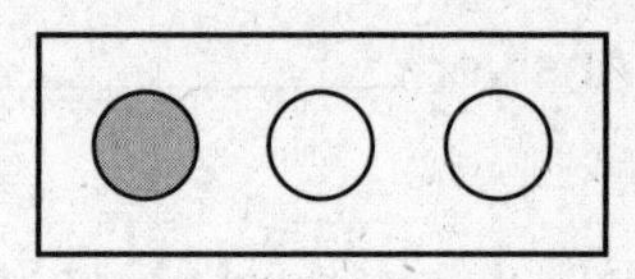

3. Write these numbers in order from least to greatest.

273	192	154	337	230

______ ______ ______ ______ ______
least   greatest

4. Use the pictograph to answer these questions.

   How many children chose purple? ______

   How many children chose yellow? ______

   How many more children chose purple than yellow? ______

   Draw pictures to show that 4 children chose orange.

Favorite Colors

Yellow	☺ ☺ ☺ ◖
Purple	☺ ☺ ☺ ☺ ☺
Orange	

☺ = 2 children

5. It's morning. What time is it?

   ____

   It's afternoon. What time is it?

   ____

6. Find the sums.

$$\begin{array}{r} 27 \\ +\ 53 \\ \hline \end{array} \qquad \begin{array}{r} 96 \\ +\ 51 \\ \hline \end{array} \qquad \begin{array}{r} 73 \\ 14 \\ +\ 23 \\ \hline \end{array} \qquad \begin{array}{r} 13 \\ 38 \\ +\ 37 \\ \hline \end{array}$$

Name ________________________________

Class Fact Practice 84A

Saxon Math 2 (for use with Lesson 84)

Set 16: Subtracting 4 and 3

Do –4 Wrap-Up twice.
Do –3 Wrap-Up twice.

$7 - 4$	$5 - 3$	$11 - 4$	$7 - 3$	$6 - 4$
$12 - 4$	$8 - 3$	$4 - 4$	$6 - 3$	$13 - 4$
$10 - 3$	$5 - 4$	$9 - 3$	$3 - 3$	$10 - 4$
$4 - 3$	$12 - 3$	$9 - 4$	$11 - 3$	$8 - 4$
$10 - 3$	$8 - 4$	$8 - 3$	$11 - 4$	$5 - 3$

Name ______________________

Saxon Math 2 (for use with **Lesson 84**)

Set 16: Subtracting 4 and 3

1. Read the answers to someone.
2. Write the answers.
3. Ask someone to correct your paper. Corrected by ______________________

$6 - 4 =$	$3 - 3 =$	$8 - 3 =$	$10 - 4 =$	$13 - 4 =$
$9 - 3 =$	$8 - 4 =$	$11 - 3 =$	$4 - 3 =$	$7 - 4 =$
$6 - 3 =$	$12 - 4 =$	$10 - 3 =$	$5 - 4 =$	$7 - 3 =$
$4 - 4 =$	$12 - 3 =$	$11 - 4 =$	$5 - 3 =$	$9 - 4 =$
$11 - 4 =$	$5 - 3 =$	$8 - 3 =$	$8 - 4 =$	$10 - 3 =$

Name .
Draw a 3-inch line segment.

Date .
Draw a $3\frac{1}{2}$" line segment.

Guided Class Practice 84A

***Saxon** Math 2 (for use with **Lesson 84**)*

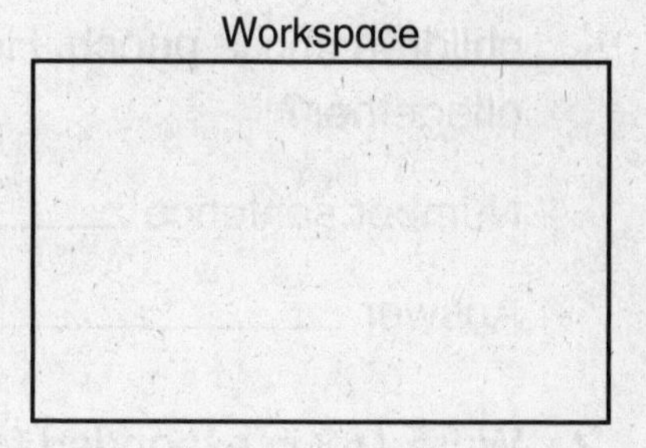

1. Forty-five children chose chocolate. Thirty-seven children chose vanilla. How many children is this altogether?

Number sentence ____________________

Answer ____________________

2. Write 437 in expanded form. ____________________

Write the number for 200 + 60 + 7. __________

Write the number for 300 + 8. __________

3. Color the thermometer to show 26°F.

4. It's afternoon. What time is it? ____________________

5. Match the numbers with the words.

105 •	• fifty-one
501 •	• one hundred fifty
150 •	• one hundred five
51 •	• five hundred one

6. What fractional part of this set of color tiles is

red? _____ blue? _____ yellow? _____

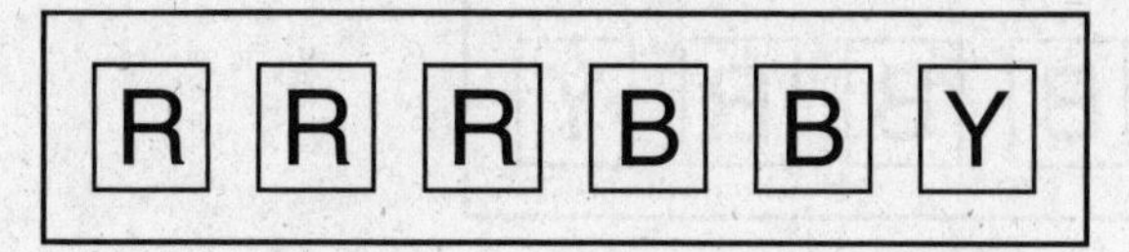

7. Find the answers.

27 + 10 = _____

48 − 10 = _____

10 more than 31 = _____

10 less than 57 = _____

6 + 7 + 5 + 3 + 4 = _____

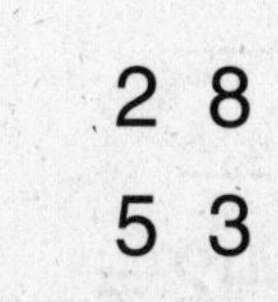

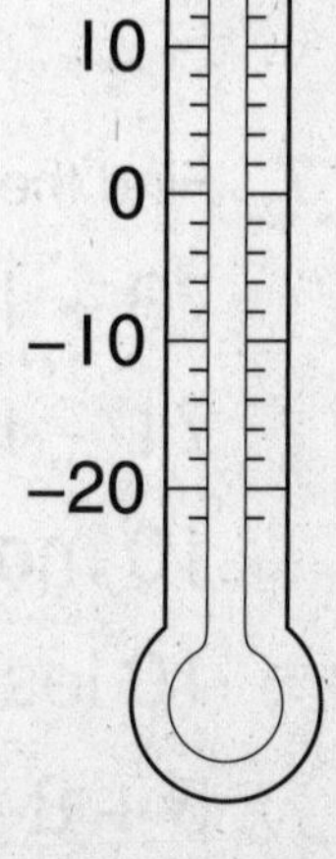

Name ______________________

Date ______________________

Workspace

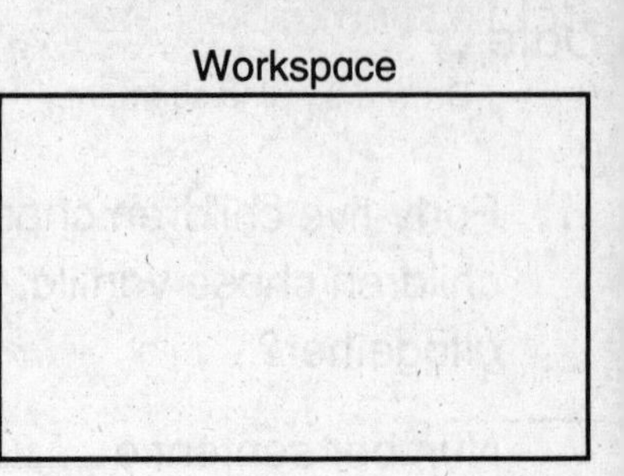

1. Twenty-eight children chose strawberry. Fifty-seven children chose peach. How many children is this altogether?

Number sentence ______________________

Answer ______________________

2. Write 163 in expanded form. ______________________

Write the number for 300 + 20 + 9. __________

Write the number for 400 + 70. __________

3. Color the thermometer to show 28°F.

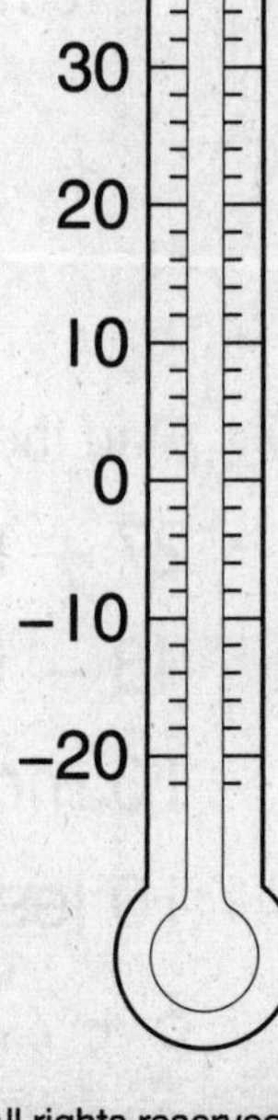

4. It's morning. What time is it? __________

5. Match the numbers with the words.

208 •	• two hundred eighty
802 •	• eight hundred two
280 •	• eighty-two
82 •	• two hundred eight

6. What fractional part of this set of color tiles is

red? ____ blue? ____ yellow? ____

7. Find the answers.

73 + 10 = ______

71 − 10 = ______

10 more than 42 = ______

10 less than 38 = ______

5 + 8 + 1 + 2 + 5 = ______

$$\begin{array}{r} 24 \\ 8 \\ +\ 82 \\ \hline \end{array}$$

Name ______________________ Score ______

Fact Assessment 16-1

***Saxon** Math 2 (for use with **Lesson 85-1**)*

Set 16: Subtracting 4 and 3

10	6	4	7	12
− 4	− 3	− 4	− 4	− 4
10	8	9	6	4
− 3	− 4	− 3	− 4	− 3
8	5	13	5	9
− 3	− 4	− 4	− 3	− 4
6	11	3	7	12
− 4	− 4	− 3	− 4	− 3
5	8	11	7	4
− 4	− 4	− 3	− 3	− 4

Name ______________________________

Fact Homework 85A

Saxon *Math 2 (for use with* ***Lesson 85-1****)*

Set 16: Subtracting 4 and 3

Pretend you are the teacher.
Correct this paper.
If the answer is incorrect, write the correct answer next to the problem.

$\begin{array}{r} 10 \\ -\ 4 \\ \hline 6 \end{array}$ $\begin{array}{r} 6 \\ -\ 3 \\ \hline 3 \end{array}$ $\begin{array}{r} 4 \\ -\ 4 \\ \hline 0 \end{array}$ $\begin{array}{r} 7 \\ -\ 4 \\ \hline 2 \end{array}$ $\begin{array}{r} 12 \\ -\ 4 \\ \hline 8 \end{array}$

$\begin{array}{r} 10 \\ -\ 3 \\ \hline 7 \end{array}$ $\begin{array}{r} 8 \\ -\ 4 \\ \hline 4 \end{array}$ $\begin{array}{r} 9 \\ -\ 3 \\ \hline 5 \end{array}$ $\begin{array}{r} 6 \\ -\ 4 \\ \hline 2 \end{array}$ $\begin{array}{r} 4 \\ -\ 3 \\ \hline 1 \end{array}$

$\begin{array}{r} 8 \\ -\ 3 \\ \hline 5 \end{array}$ $\begin{array}{r} 5 \\ -\ 4 \\ \hline 1 \end{array}$ $\begin{array}{r} 13 \\ -\ 4 \\ \hline 8 \end{array}$ $\begin{array}{r} 5 \\ -\ 3 \\ \hline 2 \end{array}$ $\begin{array}{r} 9 \\ -\ 4 \\ \hline 5 \end{array}$

$\begin{array}{r} 6 \\ -\ 4 \\ \hline 2 \end{array}$ $\begin{array}{r} 11 \\ -\ 4 \\ \hline 8 \end{array}$ $\begin{array}{r} 3 \\ -\ 3 \\ \hline 0 \end{array}$ $\begin{array}{r} 7 \\ -\ 4 \\ \hline 3 \end{array}$ $\begin{array}{r} 12 \\ -\ 3 \\ \hline 8 \end{array}$

$\begin{array}{r} 5 \\ -\ 4 \\ \hline 1 \end{array}$ $\begin{array}{r} 8 \\ -\ 4 \\ \hline 4 \end{array}$ $\begin{array}{r} 11 \\ -\ 3 \\ \hline 8 \end{array}$ $\begin{array}{r} 7 \\ -\ 3 \\ \hline 4 \end{array}$ $\begin{array}{r} 4 \\ -\ 4 \\ \hline 0 \end{array}$

Name ______________________________

Saxon Math 2 (for use with Lesson 85-1)

S50: 50 Subtraction Facts   Corrected by ________

1	7 − 1	10 − 4	9 − 0	5 − 4	11 − 3	8 − 2	9 − 4	6 − 3	10 − 2	6 − 1
2	2 − 0	9 − 3	7 − 2	11 − 4	4 − 2	8 − 0	4 − 1	8 − 4	9 − 2	5 − 1
3	1 − 0	7 − 3	4 − 3	12 − 4	8 − 3	10 − 3	4 − 0	6 − 2	7 − 0	11 − 2
4	1 − 1	12 − 3	2 − 1	2 − 2	3 − 1	10 − 1	6 − 0	13 − 4	5 − 2	3 − 0
5	3 − 3	9 − 1	6 − 4	7 − 4	4 − 4	3 − 2	5 − 0	5 − 3	0 − 0	8 − 1

Name ______________________________

Saxon *Math 2 (for use with* ***Lesson 85-1****)*

A. Write the answers.

5 – 5 = ______

6 – 5 = ______

7 – 5 = ______

8 – 5 = ______

9 – 5 = ______

10 – 5 = ______

11 – 5 = ______

12 – 5 = ______

13 – 5 = ______

14 – 5 = ______

15 – 5 = ______

16 – 5 = ______

B. Draw lines to connect the problems to the answers.

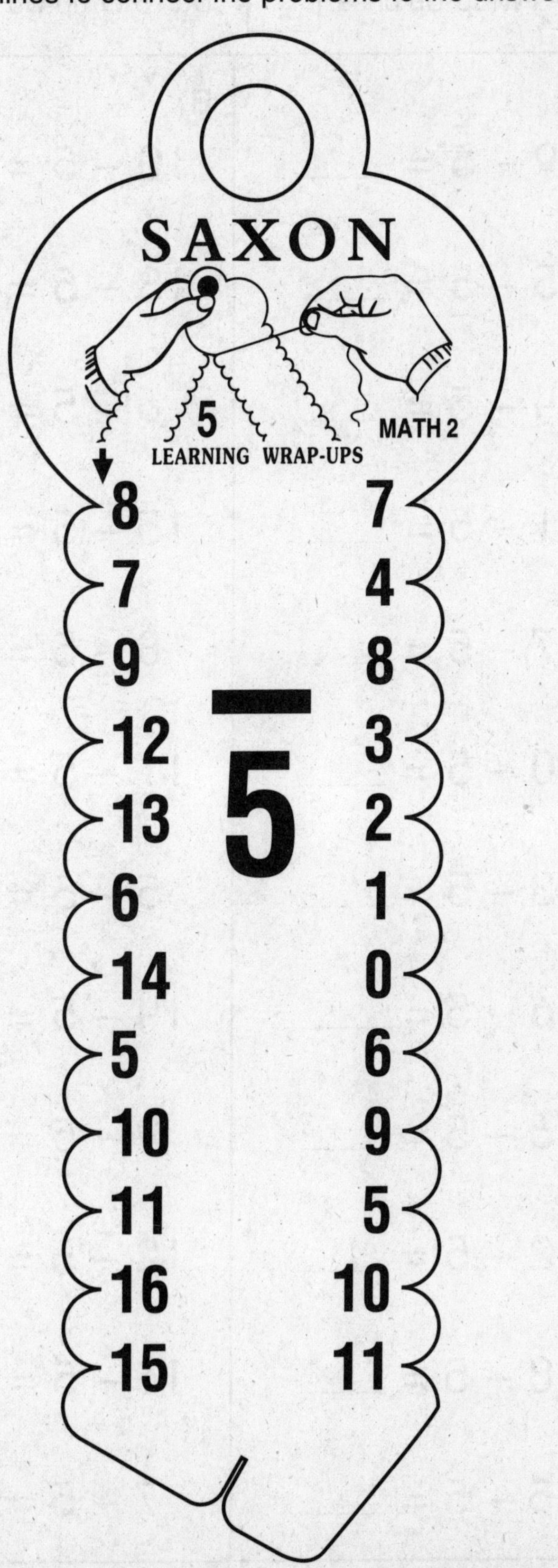

Name ____________________

Lesson Worksheet 85-1

Saxon *Math 2 (for use with* ***Lesson 85-1****)*

Do –5 Wrap-Up once. ☐	Do –5 Wrap-Up once. ☐	Do –5 Wrap-Up once. ☐
A.	**B.**	**C.**
9 – 5 = ____	10 – 5 = ____	11 – 5 = ____
5 – 5 = ____	13 – 5 = ____	13 – 5 = ____
14 – 5 = ____	6 – 5 = ____	7 – 5 = ____
11 – 5 = ____	15 – 5 = ____	9 – 5 = ____
7 – 5 = ____	8 – 5 = ____	14 – 5 = ____
10 – 5 = ____	11 – 5 = ____	6 – 5 = ____
13 – 5 = ____	5 – 5 = ____	10 – 5 = ____
6 – 5 = ____	14 – 5 = ____	15 – 5 = ____
16 – 5 = ____	9 – 5 = ____	8 – 5 = ____
12 – 5 = ____	12 – 5 = ____	12 – 5 = ____
8 – 5 = ____	16 – 5 = ____	5 – 5 = ____
15 – 5 = ____	7 – 5 = ____	16 – 5 = ____

Name
Draw a 2-inch line segment.

Date
Draw a $4\frac{1}{2}$-inch line segment.

Guided Class Practice 85A

Saxon Math 2 (for use with Lesson 85-1)

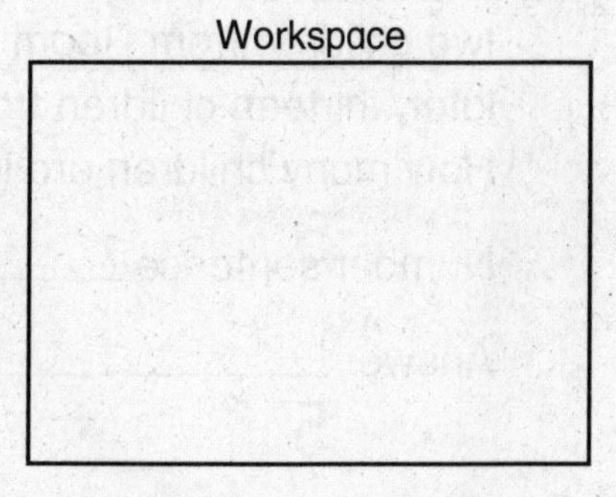

1. Fourteen children were in the gym. Twenty-five children from Room 12 joined them. Ten minutes later, fifteen children from Room 6 arrived. How many children are in the gym now?

   Number sentence ______________________

   Answer ______________________

2. One of these numbers is my secret number.
   Cross out the numbers that cannot be my secret number.

   It is not an odd number.
   It has only one digit.
   It is greater than 7.

5	6	7	8	9	10	11	12	13	14

   What is my secret number? ______

3. How much money is this? ________

4. Write these numbers in order from least to greatest.

321	114	259	170

   ______ ______ ______ ______
   least   greatest

5. Write 361 in expanded form. ______________________

6. Find the answers.

   58 − 10 = ______

   65 − 10 = ______

   $$\begin{array}{r} 58¢ \\ +\ 16¢ \\ \hline \end{array} \qquad \begin{array}{r} 17¢ \\ 53¢ \\ +\ 24¢ \\ \hline \end{array} \qquad \begin{array}{r} 26 \\ 85 \\ +\ 49 \\ \hline \end{array}$$

Name ______________________________

Date ______________________________

Homework 85B

Saxon *Math 2 (for use with* ***Lesson 85-1****)*

Workspace

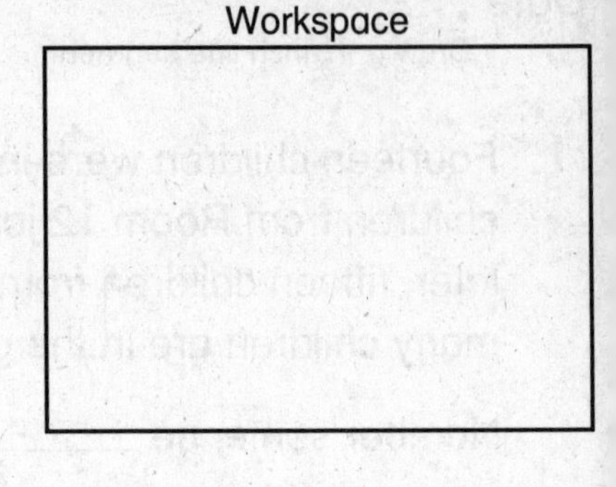

1. Eighteen children were in the lunch room. Twenty-two children from Room 7 arrived. Five minutes later, thirteen children from Room 21 joined them. How many children are in the lunch room now?

Number sentence ______________________________

Answer ______________________________

2. One of these numbers is my secret number. Cross out the numbers that cannot be my secret number.

It is not an even number.
It has two digits.
It is less than 12.

5	6	7	8	9	10	11	12	13	14

What is my secret number? ________

3. How much money is this? ____________

4. Write these numbers in order from least to greatest.

148	220	171	205

________ ________ ________ ________
least ... greatest

5. Write 432 in expanded form. ______________________________

6. Find the answers.

76 – 10 = ________

23 – 10 = ________

$$\begin{array}{r} 61¢ \\ +\ 19¢ \\ \hline \end{array} \qquad \begin{array}{r} 24¢ \\ 15¢ \\ +\ 42¢ \\ \hline \end{array} \qquad \begin{array}{r} 85 \\ 57 \\ +\ 28 \\ \hline \end{array}$$

Name ______________________ Score ______

Fact Assessment **16-2**

***Saxon** Math 2 (for use with **Lesson 85-2**)*

S50: 50 Subtraction Facts

$7 - 1$ $10 - 4$ $9 - 0$ $5 - 4$ $11 - 3$ $8 - 2$ $9 - 4$ $6 - 3$ $10 - 2$ $6 - 1$

$2 - 0$ $9 - 3$ $7 - 2$ $11 - 4$ $4 - 2$ $8 - 0$ $4 - 1$ $8 - 4$ $9 - 2$ $5 - 1$

$1 - 0$ $7 - 3$ $4 - 3$ $12 - 4$ $8 - 3$ $10 - 3$ $4 - 0$ $6 - 2$ $7 - 0$ $11 - 2$

$1 - 1$ $12 - 3$ $2 - 1$ $2 - 2$ $3 - 1$ $10 - 1$ $6 - 0$ $13 - 4$ $5 - 2$ $3 - 0$

$3 - 3$ $9 - 1$ $6 - 4$ $7 - 4$ $4 - 4$ $3 - 2$ $5 - 0$ $5 - 3$ $0 - 0$ $8 - 1$

Name ______________________

Date ______________________

Saxon *Math 2 (for use with* ***Lesson 85-2****)*

1. Amanda bought a ruler for 27¢ and an eraser for 46¢.

How much money did she spend?

Number sentence ______________________

Answer ______________________

Workspace

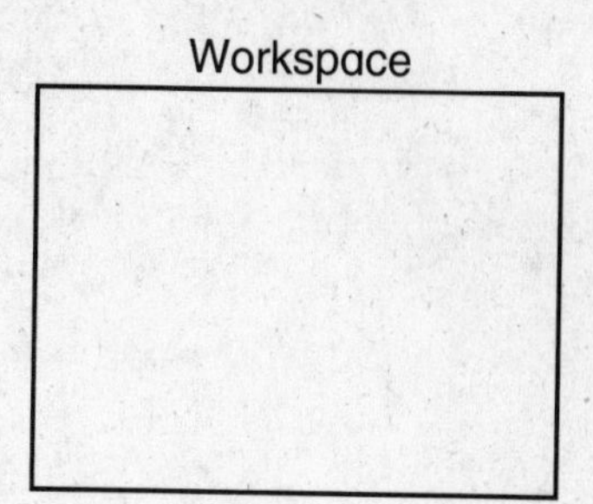

2. Color the thermometer to show 74°F.

°F

110
100
90
80
70
60
50
40
30
20
10
0
–10
–20

3. Write these numbers in order from least to greatest.

142	316	221	164	79

____ ____ ____ ____ ____

least — greatest

4. Show seven fifteen on the clocks.

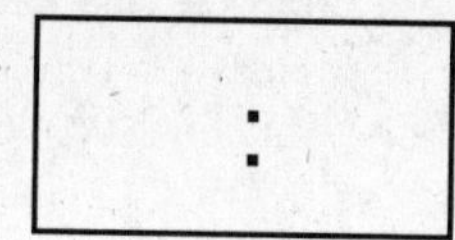

5. Find the answers.

$53 + 10 =$ ____

$26 - 10 =$ ____

$6 + 3 + 7 =$ ____

10 more than 82 = ____

one less than 30 = ____

10 less than 63 = ____

6. Find the answers.

$$\begin{array}{r} 65 \\ +\ 48 \\ \hline \end{array}$$

62¢ + 18¢

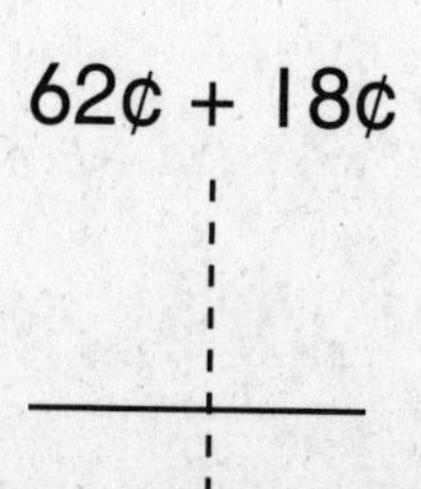

$$\begin{array}{r} 14¢ \\ 23¢ \\ +\ 36¢ \\ \hline \end{array}$$

Name ______________________________

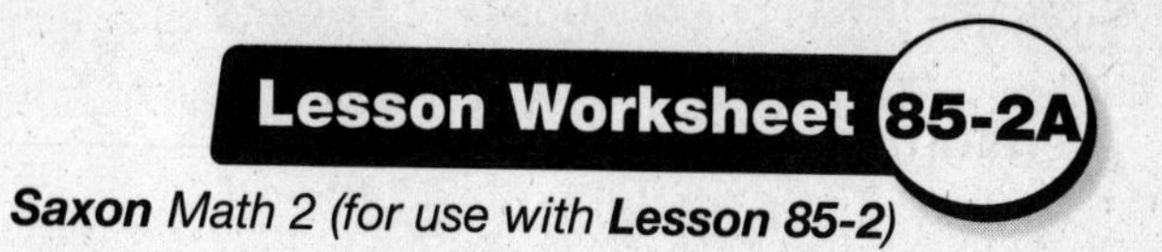

Saxon *Math 2 (for use with* ***Lesson 85-2****)*

Cover all of these shapes using only one set of tangram pieces.

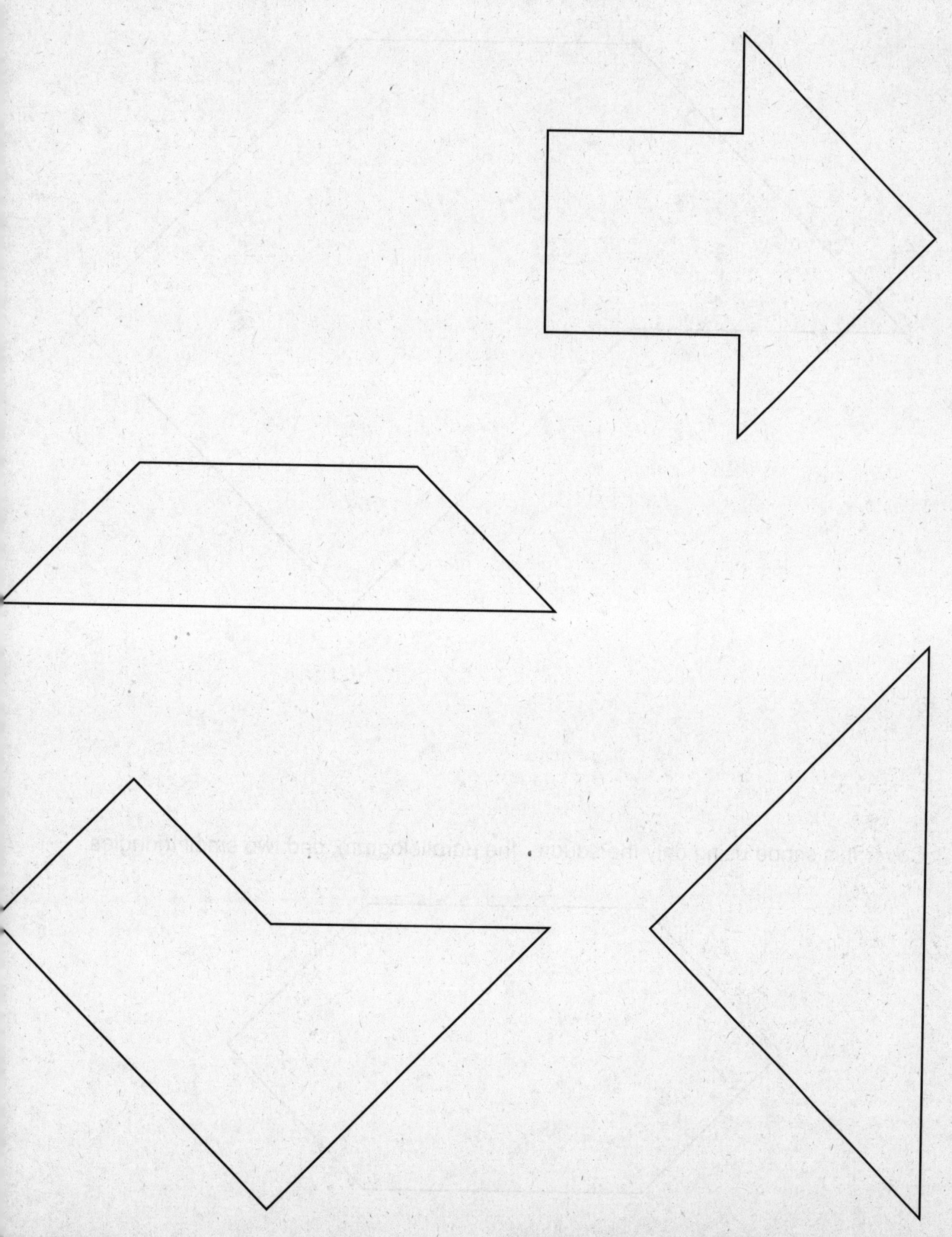

Name ___________________________

Cover this shape using only the triangles.

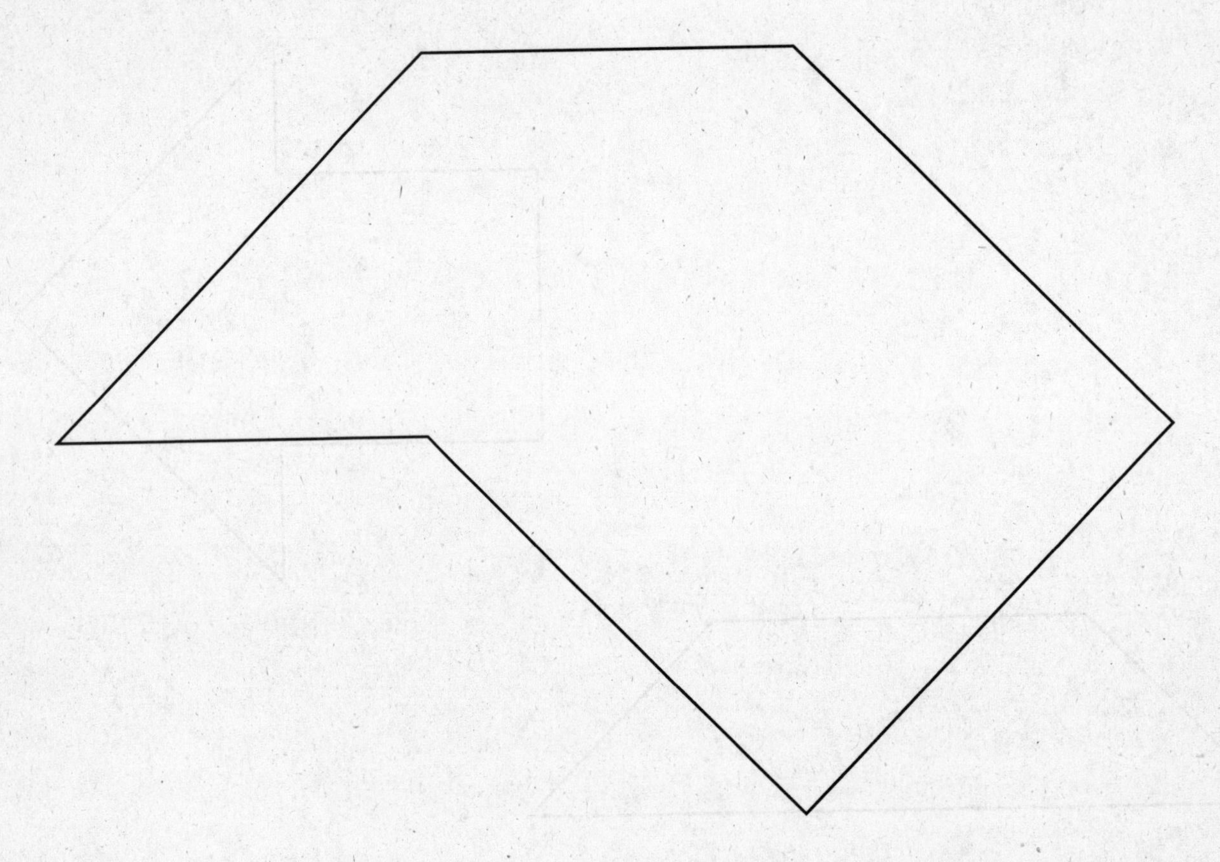

Cover this shape using only the square, the parallelogram, and two small triangles.

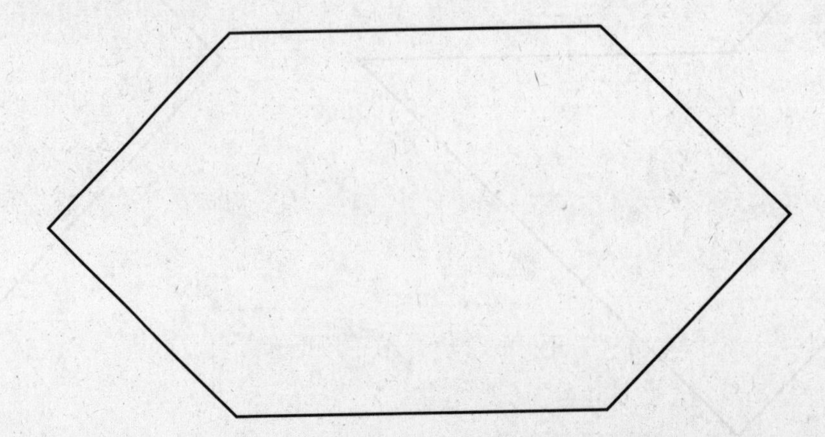

Name ____________________

Class Fact Practice 86A

***Saxon** Math 2 (for use with **Lesson 86**)*

Set 17: Subtracting 5 and 4; Review Facts

Do –5 Wrap-Up once.
Do –4 Wrap-Up once.

10 − 5	5 − 5	14 − 5	12 − 5	7 − 5
9 − 5	13 − 5	6 − 5	11 − 5	8 − 5
5 − 4	12 − 4	9 − 4	4 − 4	8 − 4
11 − 4	6 − 4	13 − 4	10 − 4	7 − 4
12 − 3	9 − 3	4 − 3	5 − 3	7 − 3

Name ______________________

Set 17: Subtracting 5 and 4; Review Facts

1. Read the answers to someone.
2. Write the answers.
3. Ask someone to correct your paper. Corrected by ______________________

11 − 5	9 − 5	14 − 5	6 − 5	12 − 5
8 − 5	10 − 5	5 − 5	13 − 5	7 − 5
10 − 4	6 − 4	13 − 4	5 − 4	7 − 4
8 − 4	4 − 4	11 − 4	9 − 4	12 − 4
7 − 3	3 − 3	12 − 3	6 − 3	10 − 3

Name . Draw a $2\frac{1}{2}$" line segment.

Date . Draw a 3" line segment.

Guided Class Practice 86A

Saxon Math 2 (for use with **Lesson 86**)

1. The kindergarten children made a graph to show the shoes and sneakers they were wearing. They counted 10 shoes and 14 sneakers. Draw a picture to show the shoes and sneakers. Circle the pairs.

How many children are in the kindergarten class? ________

2. What number does this picture show? ________
Circle the digit in the hundreds' place.
Write the number in expanded form.

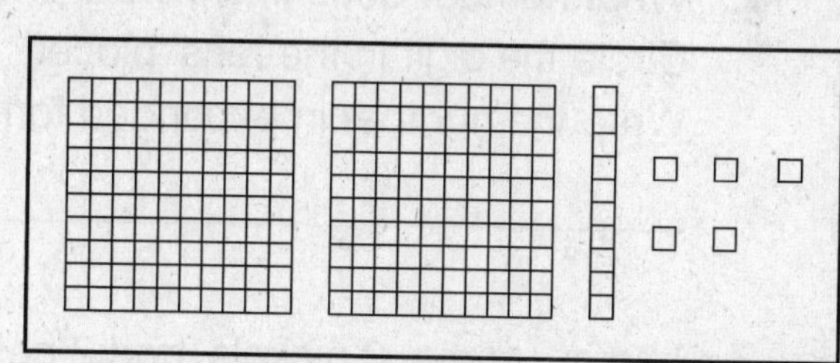

____

3. I have 3 dimes, 2 nickels, and 1 penny. Draw the coins. How much money is this? Write the amount two ways.

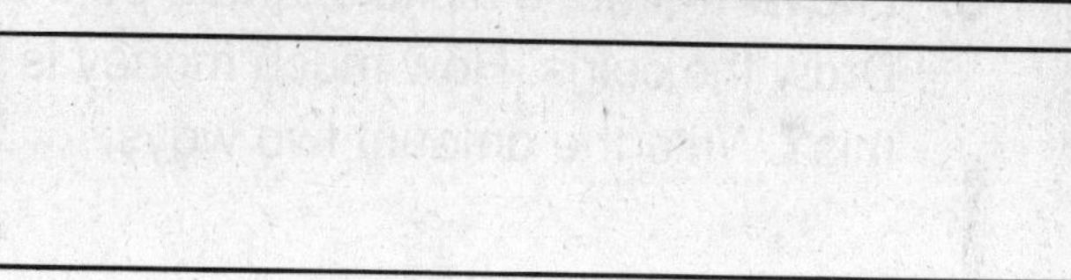

________ ________

4. Calvin has four markers. Draw the markers. One marker is red. Color that marker.

What fractional part of the markers is red? ________

5. Fill in the correct comparison symbol (>, <, or =).

4 + 2 ◯ 7 − 1 4 + 4 ◯ 4 + 5 7 − 2 ◯ 4

6. Find the sums.

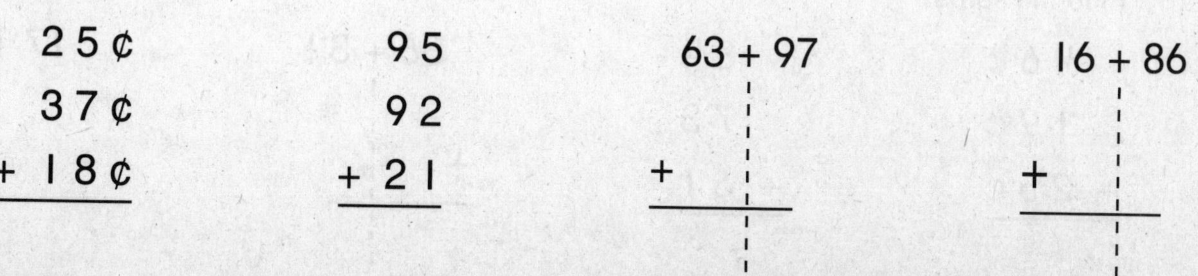

Name ______________________

Date ______________________

***Saxon** Math 2 (for use with **Lesson 86**)*

1. The first grade children made a graph to show the shoes and sneakers they were wearing. They counted 8 shoes and 12 sneakers. Draw a picture to show the shoes and sneakers. Circle the pairs.

How many children are in the first grade class? ________

2. What number does this picture show? ________
Circle the digit in the tens' place.
Write the number in expanded form.

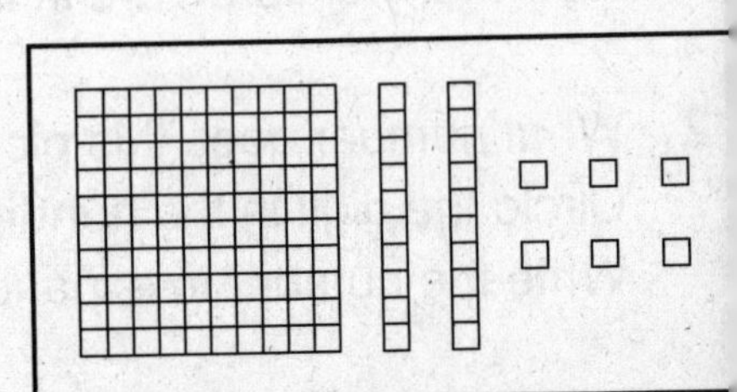

____

3. I have 1 dime, 3 nickels, and 2 pennies.
Draw the coins. How much money is this? Write the amount two ways.

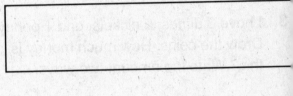

________ ________

4. Steve has three markers. Draw the markers.
One marker is green. Color that marker.

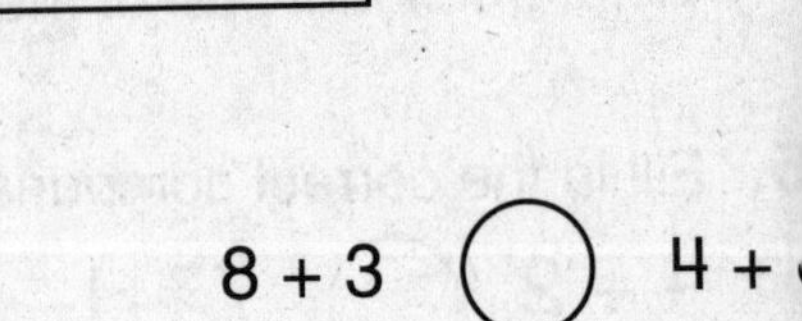

What fractional part of the markers is green? ________

5. Fill in the correct comparison symbol (>, <, or =).

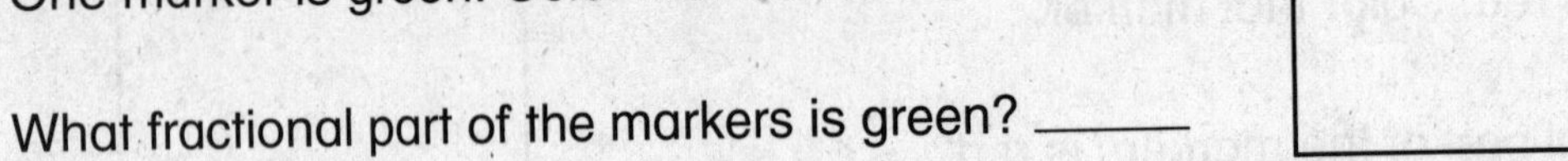

4 − 3 ◯ 7 − 5 6 + 6 ◯ 9 + 3 8 + 3 ◯ 4 +

6. Find the sums.

$$\begin{array}{r} 46¢ \\ 19¢ \\ +\ 25¢ \\ \hline \end{array} \qquad \begin{array}{r} 84 \\ 73 \\ +\ 61 \\ \hline \end{array}$$

86 + 34 17 + 9

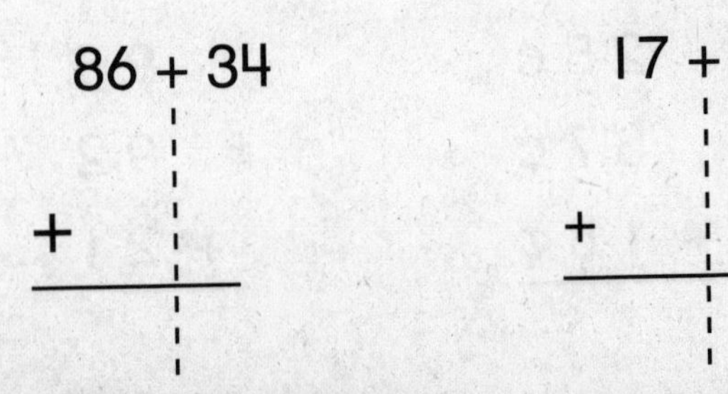

Name ______________________

Class Fact Practice **87A**

Saxon *Math 2 (for use with* ***Lesson 87****)*

Set 17: Subtracting 5 and 4

Do –5 Wrap-Up twice. ☐☐ Do –4 Wrap-Up twice. ☐☐

7 – 5 = ____	8 – 4 = ____
10 – 5 = ____	4 – 4 = ____
14 – 5 = ____	13 – 4 = ____
5 – 5 = ____	5 – 4 = ____
9 – 5 = ____	10 – 4 = ____
12 – 5 = ____	7 – 4 = ____
6 – 5 = ____	12 – 4 = ____
13 – 5 = ____	6 – 4 = ____
8 – 5 = ____	11 – 4 = ____
11 – 5 = ____	9 – 4 = ____

Name ______________________

Set 17: Subtracting 5 and 4

1. Write the answers.
2. Draw lines to connect the problems that have the same answer.
3. Ask someone to correct your paper. Corrected by ______________________

6 – 5 = ____ •	• 10 – 4 = ____
13 – 5 = ____ •	• 6 – 4 = ____
5 – 5 = ____ •	• 9 – 4 = ____
11 – 5 = ____ •	• 5 – 4 = ____
9 – 5 = ____ •	• 12 – 4 = ____
7 – 5 = ____ •	• 4 – 4 = ____
10 – 5 = ____ •	• 8 – 4 = ____
12 – 5 = ____ •	• 13 – 4 = ____
8 – 5 = ____ •	• 11 – 4 = ____
14 – 5 = ____ •	• 7 – 4 = ____

Name ____________________

Measure this line segment using inches. ______"

Date

Draw a $1\frac{1}{2}$" line segment.

Guided Class Practice 87A

***Saxon** Math 2 (for use with **Lesson 87**)*

1. There are 83 children in Grade 2 at Haley School. Ten second graders were absent on Monday. How many Grade 2 children were in school?

Number sentence ____________________

Answer ____________________

2. Shelly has a half-dozen dimes and a dozen pennies.

How many dimes is this? ______ How much money is that? ______

How many pennies is this? ______ How much money is that? ______

Write how much money Shelly has in two ways. ______ ______

3. Fill in the missing numbers in this number pattern.

______, ______, ______, 35, 40, 45, ______, ______, ______

4. It's evening. What time is it? ____________

5. Draw a line of symmetry in each shape.

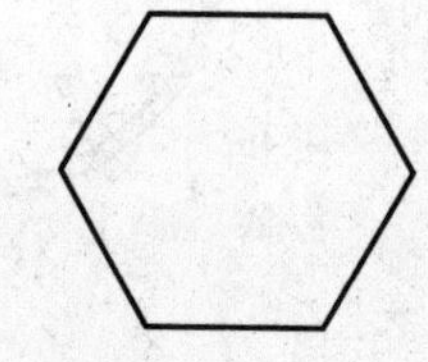

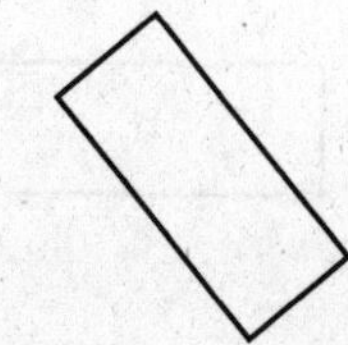

6. Color $\frac{2}{3}$ of the flowers.

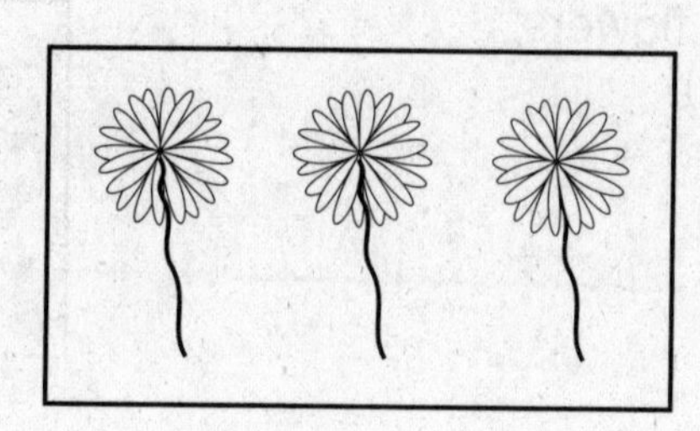

7. Find the sums.

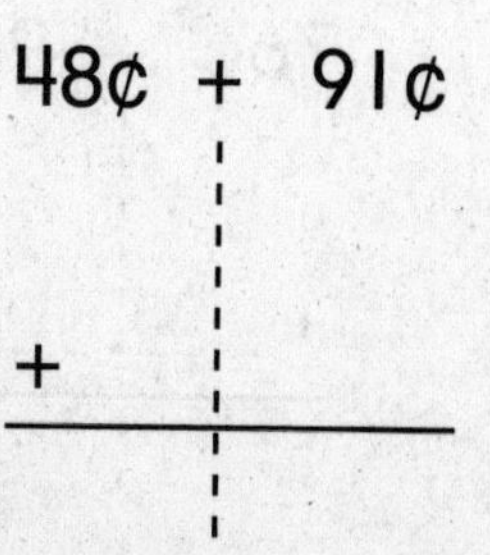

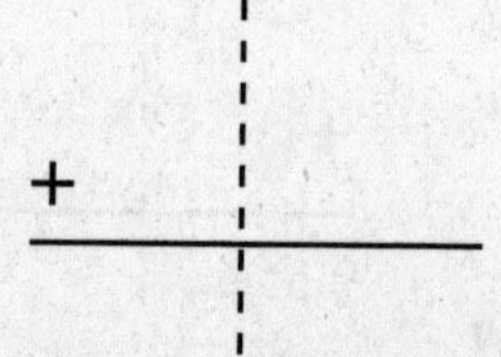

87¢ + 84¢

+

Name ______________________

Date ______________________

Saxon Math 2 (for use with **Lesson 87**)

1. The label on the bag says that there are 72 pieces of candy in the bag. Mindy ate 10 pieces. How many pieces are left?

Number sentence ______________________

Answer ______________________

2. Curtis has nine dimes and a half-dozen pennies.

How many dimes is this? ________ How much money is that? ________

How many pennies is this? ________ How much money is that? ________

Write how much money Curtis has in two ways. ________ ________

3. Fill in the missing numbers in this number pattern.

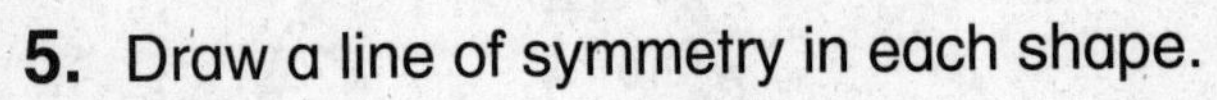

______, ______, ______, 50, 45, 40, ______, ______, ______

4. It's morning. What time is it? ______________

5. Draw a line of symmetry in each shape.

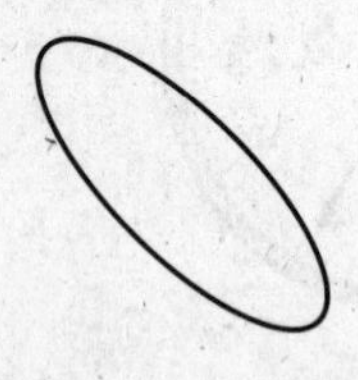

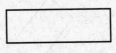

6. Color $\frac{1}{3}$ of the flowers.

7. Find the sums.

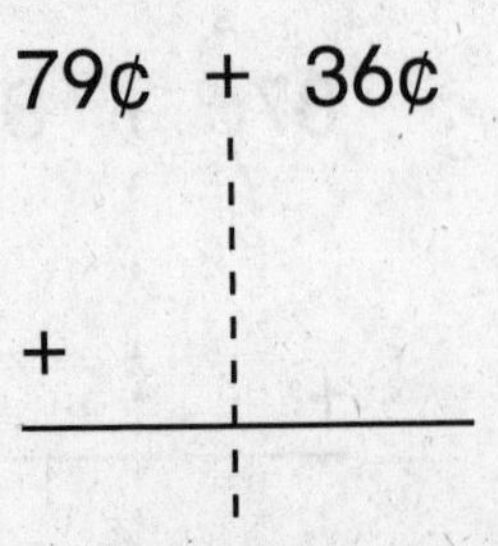

79¢ + 36¢ 43¢ + 82¢ 76¢ + 94¢

Name ______________________________

Class Fact Practice **88A**

***Saxon** Math 2 (for use with **Lesson 88**)*

Set 17: Subtracting 5 and 4

10 − 5	5 − 4	13 − 5	5 − 5	7 − 4
12 − 5	10 − 4	6 − 5	8 − 4	11 − 5
9 − 4	14 − 5	11 − 4	7 − 5	13 − 4
11 − 5	8 − 5	6 − 4	12 − 5	10 − 5
4 − 4	13 − 5	9 − 5	12 − 4	14 − 5

Name ____________________

Saxon Math 2 (for use with Lesson 88)

Set 17: Subtracting 5 and 4

1. Read the answers to someone.
2. Write the answers.
3. Ask someone to correct your paper. Corrected by ____________________

7 − 4	5 − 5	13 − 5	5 − 4	10 − 5
11 − 5	8 − 4	6 − 5	10 − 4	12 − 5
13 − 4	7 − 5	11 − 4	14 − 5	9 − 4
10 − 5	12 − 5	6 − 4	8 − 5	11 − 5
14 − 5	12 − 4	9 − 5	13 − 5	4 − 4

Name

Measure this line segment using inches. ______ "

Date

Draw a 2" line segment.

1. The chart shows the number of baseball cards the children have in their collections.

Baseball Cards

Cathy	164
Steve	248
Paul	187
Ben	128

Who has the most cards? ______________

Who has the fewest cards? ______________

Write the names of the children in order from the one who has the most cards to the one who has the fewest cards.

______ most ______ ______ ______ fewest

2. Draw a picture to show three hundred twenty-one.

(Use ☐ for 100, ▯ for 10, and ▫ for 1.)

Write this number in expanded form. ______________

Circle the number that shows three hundred twenty-one.

3002001 321 30021 3021

3. Shelley has these coins. Write this money amount in two ways. ______ ______

4. Divide this square into fourths using only horizontal line segments.

Shade $\frac{2}{4}$ of the square.

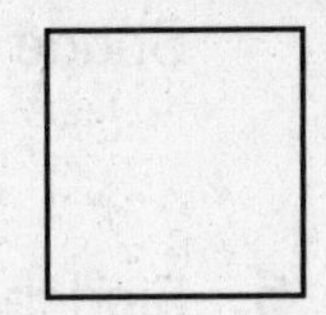

5. Find the answers.

$$\begin{array}{r} 58¢ \\ +\ 23¢ \\ \hline \end{array}$$

$$\begin{array}{r} 29¢ \\ 31¢ \\ +\ 24¢ \\ \hline \end{array}$$

54 − 10 = ______

4 + 3 + 2 + 7 = ______

Name ______________________

Date ______________________

Saxon *Math 2 (for use with* ***Lesson 88****)*

1. The chart shows the number of stamps the children have in their collections.

Who has the most stamps? ______________

Who has the fewest stamps? ______________

Stamps	
Michael	259
Vera	145
Crystal	232
Selby	95

Write the names of the children in order from the one who has the most stamps to the one who has the fewest stamps.

______ most ______ ______ ______ fewest

2. Draw a picture to show two hundred fifty-three.

(Use □ for 100, ▯ for 10, and ▫ for 1.)

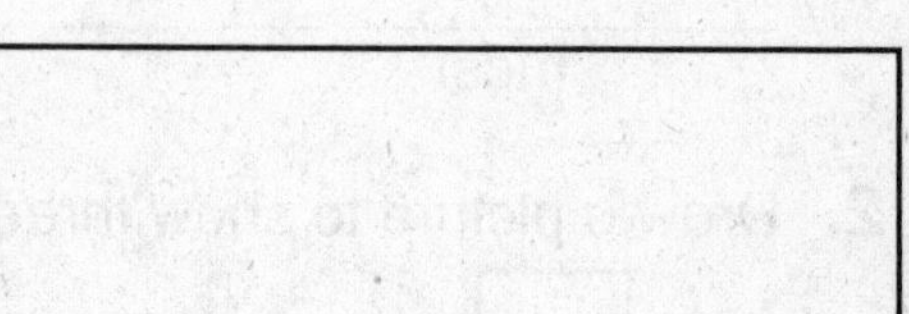

Write this number in expanded form. ______________

Circle the number that shows two hundred fifty-three.

20053 2053 200503 253

3. Steve has these coins. Write this money amount in two ways. ______ ______

4. Divide this square into fourths using only vertical line segments.

Shade $\frac{3}{4}$ of the square.

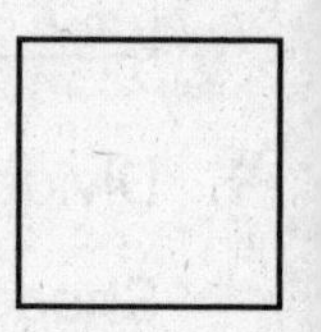

5. Find the answers.

$$\begin{array}{r} 24¢ \\ +\ 57¢ \\ \hline \end{array}$$

$$\begin{array}{r} 32¢ \\ 14¢ \\ +\ 27¢ \\ \hline \end{array}$$

42 − 10 = ______

7 + 4 + 2 + 6 = ______

Name ______________________________

Set 17: Subtracting 5 and 4

Do −5 Wrap-Up twice.
Do −4 Wrap-Up once.

8 − 5	11 − 4	14 − 5	9 − 4	7 − 5
6 − 4	10 − 5	12 − 4	5 − 5	7 − 4
11 − 5	5 − 4	13 − 5	10 − 4	9 − 5
8 − 4	6 − 5	13 − 4	4 − 4	12 − 5
7 − 5	8 − 4	9 − 4	12 − 5	10 − 4

Name ____________________________

Set 17: Subtracting 5 and 4

1. Read the answers to someone.
2. Write the answers.
3. Ask someone to correct your paper. Corrected by ____________________

$9 - 5$	$5 - 4$	$14 - 5$	$5 - 5$	$9 - 4$
$13 - 5$	$7 - 4$	$6 - 5$	$13 - 4$	$10 - 5$
$8 - 5$	$12 - 4$	$4 - 4$	$11 - 5$	$6 - 4$
$10 - 4$	$12 - 5$	$8 - 4$	$7 - 5$	$11 - 4$
$9 - 4$	$14 - 5$	$10 - 4$	$6 - 5$	$8 - 4$

Name

Draw a $2\frac{1}{2}$-inch line segment.

Date

Measure this line segment using inches. ______"

1. There are 17 children in Room 12. Ten children went to the nurse's office to have their eyes checked. How many children are in Room 12 now?

Number sentence ______________________________

Answer ____________________

2. Shelton has 15 markers, Hector has 23 markers, and Tyler has 34 markers. How many markers do the three boys have altogether?

Workspace

Number sentence ____________________

Answer ____________________

3. Use the clues to write the children's names on the Venn diagram.

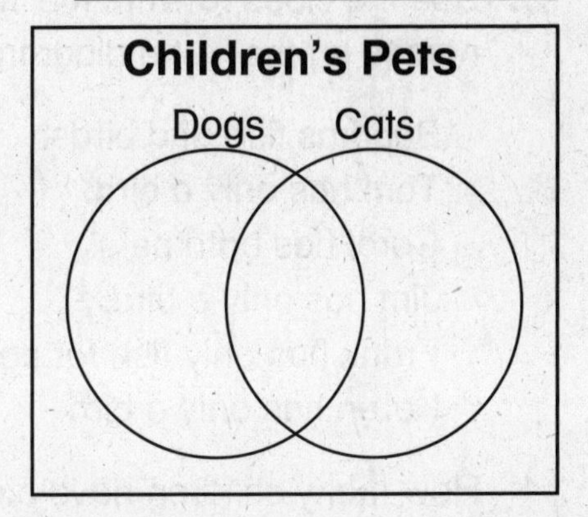

Sue has only a dog.
Mary has a cat and a dog.
Peter has only a cat.
Joe has both pets.
Sam has only a cat.

How many children have a cat? ______

How many children have only a dog? ______

4. What number does this picture show? ______
Circle the digit in the tens' place.

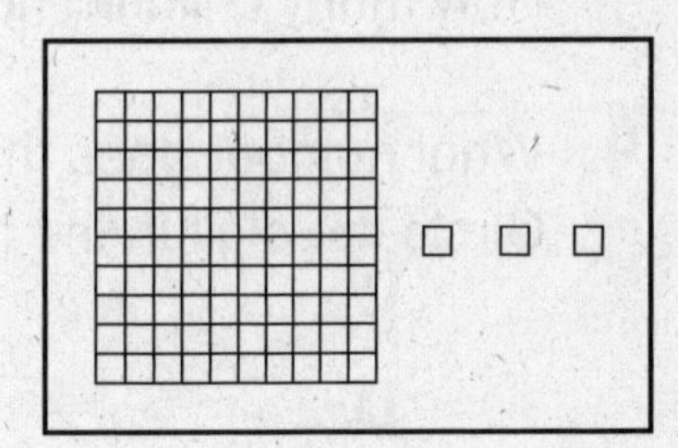

5. Fill in a number to make each number sentence true.

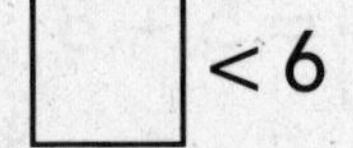 < 6 $\qquad$ $13 < \square$ $\qquad$ $4 + 7 = 7 + \square$

6. Find the answers.

10 less than 54 = ______ $\qquad$ 10 more than 31 = ______

$8 + 3 + 9 + 2 + 6 + 1 + 7 + 4 =$ ______

$$\begin{array}{r} 63¢ \\ +\ 18¢ \\ \hline \end{array}$$

Name ________________________

Date ________________________

Saxon Math 2 (for use with Lesson 89)

1. There are 25 children in Room 17. Ten children went to the library. How many children are in Room 17 now?

Number sentence ______________________________

Answer ____________________

2. Courtney has 37 markers, Sandra has 28 markers, and Maggie has 21 markers. How many markers do the three girls have altogether?

Number sentence ______________________

Answer ____________________

Workspace

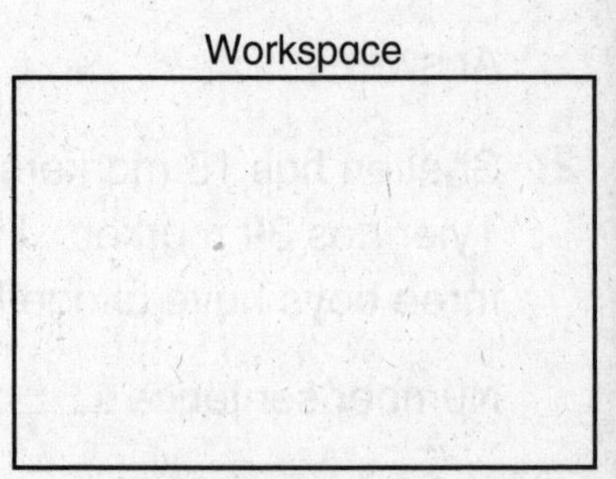

3. Use the clues to write the children's names on the Venn diagram.

Bob has fish and birds.
Tom has only a bird.
Carol has both pets.
Jim has only a bird.
Frank has only fish for pets.
Karen has only a bird.

Children's Pets

Fish Birds

How many children have birds? ______

How many children have only fish? ______

4. What number does this picture show? ______
Circle the digit in the tens' place.

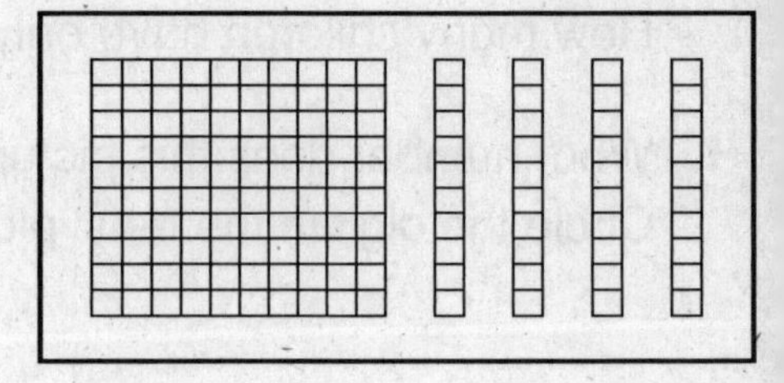

5. Fill in a number to make each number sentence true.

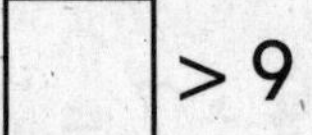 > 9 $7 >$ ☐ $8 +$ ☐ $= 5 + 8$

6. Find the answers.

10 more than 63 = ______   10 less than 41 = ______

$6 + 7 + 3 + 2 + 4 + 5 + 1 =$ ______

$$\begin{array}{r} 47¢ \\ +\ 23¢ \\ \hline \end{array}$$

Name ____________________ Score ________

Fact Assessment 17-1

***Saxon** Math 2 (for use with **Lesson 90-1**)*

Set 17: Subtracting 5 and 4

10	7	6	4	9
− 5	− 4	− 5	− 4	− 5
11	8	12	8	7
− 4	− 5	− 5	− 4	− 5
10	13	5	9	11
− 4	− 5	− 5	− 4	− 5
5	7	13	6	8
− 4	− 5	− 4	− 5	− 5
14	6	9	12	5
− 5	− 4	− 5	− 4	− 5

Name ______________________________

Fact Homework 90A

Saxon *Math 2 (for use with* ***Lesson 90-1****)*

Set 17: Subtracting 5 and 4

Pretend you are the teacher.
Correct this paper.
If the answer is incorrect, write the correct answer next to the problem.

$\begin{array}{r} 10 \\ -\ 5 \\ \hline 5 \end{array}$	$\begin{array}{r} 7 \\ -\ 4 \\ \hline 3 \end{array}$	$\begin{array}{r} 6 \\ -\ 5 \\ \hline 1 \end{array}$	$\begin{array}{r} 4 \\ -\ 4 \\ \hline 0 \end{array}$	$\begin{array}{r} 9 \\ -\ 5 \\ \hline 3 \end{array}$
$\begin{array}{r} 11 \\ -\ 4 \\ \hline 7 \end{array}$	$\begin{array}{r} 8 \\ -\ 5 \\ \hline 3 \end{array}$	$\begin{array}{r} 12 \\ -\ 5 \\ \hline 8 \end{array}$	$\begin{array}{r} 8 \\ -\ 4 \\ \hline 4 \end{array}$	$\begin{array}{r} 7 \\ -\ 5 \\ \hline 2 \end{array}$
$\begin{array}{r} 10 \\ -\ 4 \\ \hline 6 \end{array}$	$\begin{array}{r} 13 \\ -\ 5 \\ \hline 9 \end{array}$	$\begin{array}{r} 5 \\ -\ 5 \\ \hline 0 \end{array}$	$\begin{array}{r} 9 \\ -\ 4 \\ \hline 5 \end{array}$	$\begin{array}{r} 11 \\ -\ 5 \\ \hline 6 \end{array}$
$\begin{array}{r} 5 \\ -\ 4 \\ \hline 1 \end{array}$	$\begin{array}{r} 7 \\ -\ 5 \\ \hline 2 \end{array}$	$\begin{array}{r} 13 \\ -\ 4 \\ \hline 8 \end{array}$	$\begin{array}{r} 6 \\ -\ 5 \\ \hline 1 \end{array}$	$\begin{array}{r} 8 \\ -\ 5 \\ \hline 3 \end{array}$
$\begin{array}{r} 14 \\ -\ 5 \\ \hline 9 \end{array}$	$\begin{array}{r} 6 \\ -\ 4 \\ \hline 2 \end{array}$	$\begin{array}{r} 9 \\ -\ 5 \\ \hline 4 \end{array}$	$\begin{array}{r} 12 \\ -\ 4 \\ \hline 8 \end{array}$	$\begin{array}{r} 5 \\ -\ 5 \\ \hline 0 \end{array}$

Name ____________________

Saxon Math 2 (for use with Lesson 90-1)

S60: 60 Subtraction Facts   Corrected by ________

1	7 − 1	10 − 4	9 − 0	5 − 4	11 − 3	8 − 2	9 − 4	6 − 3	10 − 2	6 − 1
2	2 − 0	9 − 3	7 − 2	6 − 5	5 − 5	11 − 4	4 − 2	8 − 0	14 − 5	4 − 0
3	9 − 5	8 − 4	9 − 2	11 − 5	5 − 1	8 − 5	1 − 0	7 − 3	4 − 3	10 − 5
4	12 − 4	8 − 3	10 − 3	4 − 1	6 − 2	13 − 5	7 − 0	11 − 2	1 − 1	12 − 3
5	7 − 5	2 − 1	2 − 2	12 − 5	3 − 1	10 − 1	6 − 0	13 − 4	5 − 2	3 − 0
6	3 − 3	9 − 1	6 − 4	7 − 4	4 − 4	3 − 2	5 − 0	5 − 3	0 − 0	8 − 1

Name ______________________________

Lesson Worksheet 90-1A

***Saxon** Math 2 (for use with **Lesson 90-1**)*

A. Write the answers.

$6 - 6 =$ ______

$7 - 6 =$ ______

$8 - 6 =$ ______

$9 - 6 =$ ______

$10 - 6 =$ ______

$11 - 6 =$ ______

$12 - 6 =$ ______

$13 - 6 =$ ______

$14 - 6 =$ ______

$15 - 6 =$ ______

$16 - 6 =$ ______

$17 - 6 =$ ______

B. Draw lines to connect the problems to the answers.

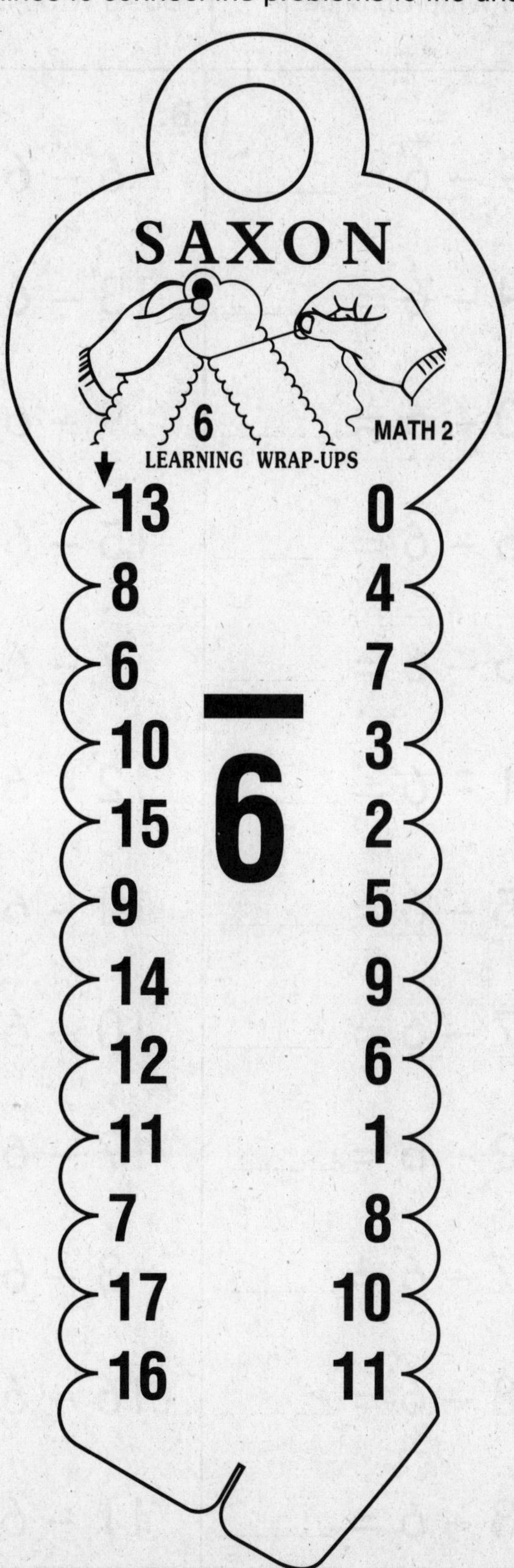

Name ______________________

*Saxon Math 2 (for use with **Lesson 90-1**)*

Do –6 Wrap-Up once. ☐	Do –6 Wrap-Up once. ☐	Do –6 Wrap-Up once. ☐
A.	**B.**	**C.**
9 – 6 = ____	6 – 6 = ____	13 – 6 = ____
14 – 6 = ____	13 – 6 = ____	8 – 6 = ____
10 – 6 = ____	9 – 6 = ____	12 – 6 = ____
16 – 6 = ____	15 – 6 = ____	9 – 6 = ____
6 – 6 = ____	7 – 6 = ____	15 – 6 = ____
11 – 6 = ____	12 – 6 = ____	6 – 6 = ____
15 – 6 = ____	14 – 6 = ____	10 – 6 = ____
7 – 6 = ____	10 – 6 = ____	16 – 6 = ____
12 – 6 = ____	17 – 6 = ____	11 – 6 = ____
17 – 6 = ____	8 – 6 = ____	14 – 6 = ____
8 – 6 = ____	16 – 6 = ____	7 – 6 = ____
13 – 6 = ____	11 – 6 = ____	17 – 6 = ____

Name ______________________

Date ______________________

Problem-Solving Worksheet 90A

***Saxon** Math 2 (for use with **Lesson 90-1**)*

Understand	Plan	Solve	Check

Guess and Check

The chart shows the prices of school supplies the children can buy at the school store.

Marker	Folder	Eraser	Ruler
37¢	46¢	23¢	27¢

Kristal has $1.00. Show three different things she can buy with her money.

What are three things Kristal can buy with her money?

______________ ______________ ______________

Name ____________________

Performance Task Worksheet 90

Saxon *Math 2 (for use with **Lesson 90-1**)*

Understand	Plan	Solve	Check

The chart shows the prices of toys the children can buy at the school fair.

Toy car	Game	Ball	Puzzle
53¢	36¢	74¢	45¢

Nathan has $1.00. Show two different things he can buy with his money.

What are two things Nathan can buy with his money?

____________________ ____________________

Circle the problem-solving strategies you used to solve this problem.

Act It Out ***Use Logical Reasoning***

Draw a Picture ***Look for a Pattern***

Make an Organized List ***Guess and Check***

Explain how you got your answer: ____________________

____

____

____

Name ______________________ Score ________

Fact Assessment 17-2

***Saxon** Math 2 (for use with **Lesson 90-2**)*

S60: 60 Subtraction Facts

1	$7 - 1$	$10 - 4$	$9 - 0$	$5 - 4$	$11 - 3$	$8 - 2$	$9 - 4$	$6 - 3$	$10 - 2$	$6 - 1$
2	$2 - 0$	$9 - 3$	$7 - 2$	$6 - 5$	$5 - 5$	$11 - 4$	$4 - 2$	$8 - 0$	$14 - 5$	$4 - 0$
3	$9 - 5$	$8 - 4$	$9 - 2$	$11 - 5$	$5 - 1$	$8 - 5$	$1 - 0$	$7 - 3$	$4 - 3$	$10 - 5$
4	$12 - 4$	$8 - 3$	$10 - 3$	$4 - 1$	$6 - 2$	$13 - 5$	$7 - 0$	$11 - 2$	$1 - 1$	$12 - 3$
5	$7 - 5$	$2 - 1$	$2 - 2$	$12 - 5$	$3 - 1$	$10 - 1$	$6 - 0$	$13 - 4$	$5 - 2$	$3 - 0$
6	$3 - 3$	$9 - 1$	$6 - 4$	$7 - 4$	$4 - 4$	$3 - 2$	$5 - 0$	$5 - 3$	$0 - 0$	$8 - 1$

Name ______________________

Date .

Draw a $3\frac{1}{2}$" line segment.

Written Assessment 17

Saxon Math 2 (for use with Lesson 90-2)

1. Sam has 18 stickers, Cedric has 27 stickers, and Tony has 32 stickers. How many stickers do the three boys have altogether?

Workspace

Number sentence ______________________

Answer ______________________

2. The chart shows how many pennies each child has.

Name	Pennies
Barbara	314
Celina	276
Amber	358
Megan	298

Who has the most pennies? ______________

Who has the fewest pennies? ______________

Write the names of the children in order from the one who has the most pennies to the one who has the fewest pennies.

________ most   ________   ________   ________ fewest

3. Measure each line segment using inches.

•————————• ______"

•——————————————————————————• ______"

4. It's evening. What time is it? ______________

5. What number does this picture show? ______

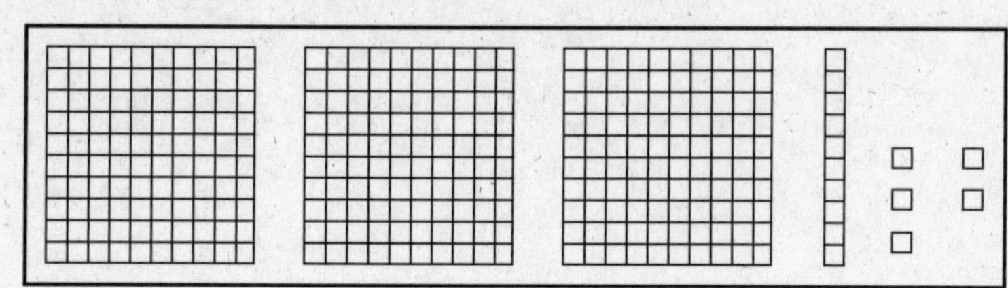

6. Find the answers.

27 – 10 = ______

58 – 10 = ______

$$\begin{array}{r} 74¢ \\ +\ 17¢ \\ \hline \end{array} \qquad \begin{array}{r} 23¢ \\ 42¢ \\ +\ 15¢ \\ \hline \end{array} \qquad \begin{array}{r} 71 \\ 19 \\ +\ 34 \\ \hline \end{array}$$

Name ______________________________

Lesson Worksheet 90-2

***Saxon** Math 2 (for use with **Lesson 90-2**)*

Cover each shape using tangram pieces.

A.

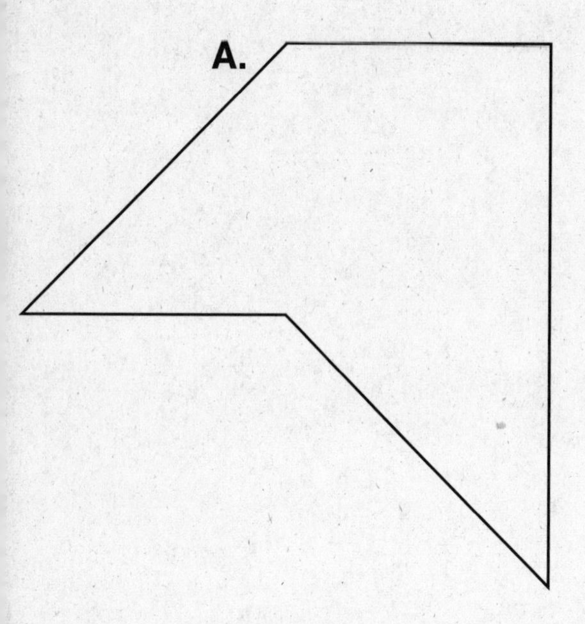

B.

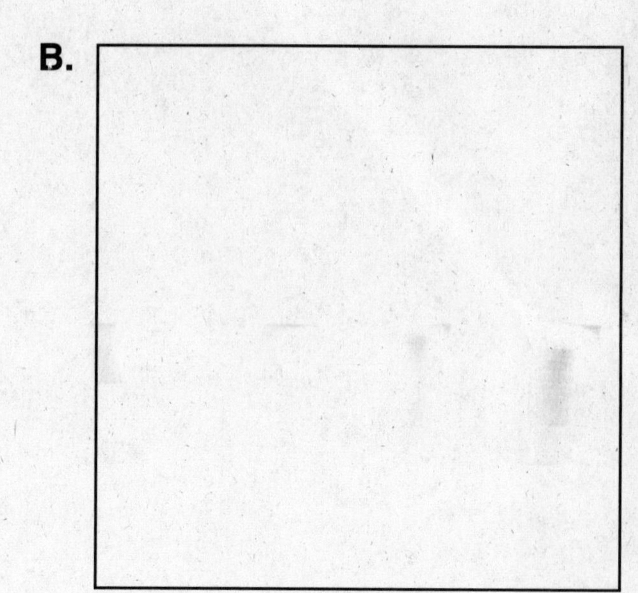

C.

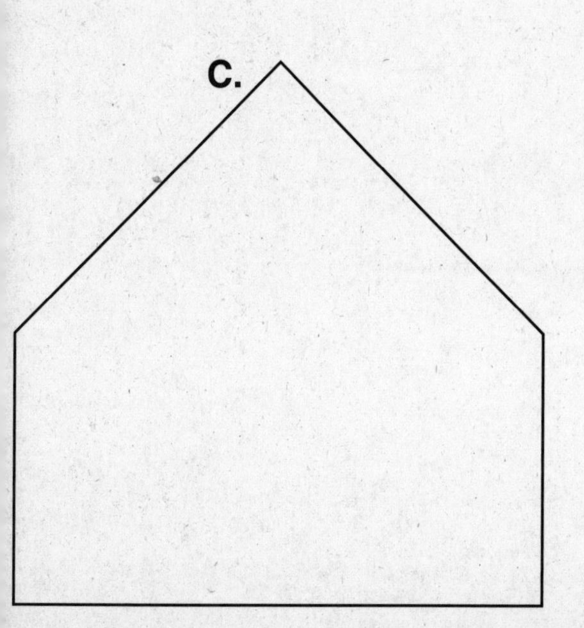

D.

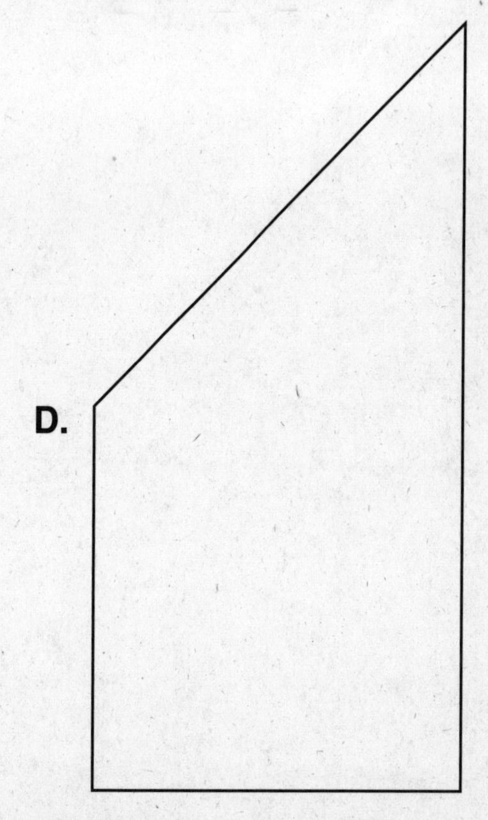

Name ______________________________

Class Fact Practice 91A

***Saxon** Math 2 (for use with **Lesson 91**)*

Set 18: Subtracting 6 and 5; Review Facts

Do −6 Wrap-Up once.
Do −5 Wrap-Up once.

8 − 6	14 − 6	12 − 6	6 − 6	11 − 6
13 − 6	9 − 6	7 − 6	15 − 6	10 − 6
9 − 5	13 − 5	5 − 5	11 − 5	6 − 5
14 − 5	8 − 5	12 − 5	7 − 5	10 − 5
12 − 4	6 − 4	5 − 4	4 − 4	10 − 4

Name ____________________

Saxon Math 2 (for use with Lesson 91)

Set 18: Subtracting 6 and 5; Review Facts

1. Read the answers to someone.
2. Write the answers.
3. Ask someone to correct your paper. Corrected by ____________________

6 − 6	12 − 6	15 − 6	8 − 6	11 − 6
10 − 6	14 − 6	7 − 6	13 − 6	9 − 6
14 − 5	9 − 5	12 − 5	5 − 5	8 − 5
10 − 5	6 − 5	11 − 5	7 − 5	13 − 5
11 − 4	5 − 4	10 − 4	7 − 4	6 − 4

Name •
Draw a $3\frac{1}{2}$" line segment.

Date •
Draw a $1\frac{1}{2}$" line segment.

1. There were a dozen stuffed animals on Roseann's bed. Her dog chewed five stuffed animals. How many stuffed animals does Roseann have left?

Number sentence ______________________

Answer ______________

2. Draw pictures to show the number of jelly beans each child has.

Child	Number of Jelly Beans
Shelley	12
Danielle	6
Joshua	9

Jelly Beans

Shelley	
Danielle	
Joshua	

⬭ = 2 jelly beans

3. Find the differences.

$$\begin{array}{r} 63¢ \\ -\ 35¢ \\ \hline \end{array} \qquad \begin{array}{r} 48¢ \\ -\ 16¢ \\ \hline \end{array} \qquad \begin{array}{r} 50¢ \\ -\ 34¢ \\ \hline \end{array}$$

4. Put these numbers in order from least to greatest.
Circle the number that is between 28 and 42.

29	43	27	48	19

_____ _____ _____ _____ _____
least greatest

5. Match the numbers with the words.

137 • • three hundred seventeen
731 • • seven hundred thirteen
317 • • one hundred thirty-seven
713 • • seven hundred thirty-one

6. Brian has 3 dimes and 16 pennies.
How much money is this?
Write the amount two ways. __________ __________

Name ______________________

Date ______________________

Homework 91B

***Saxon** Math 2 (for use with **Lesson 91**)*

1. Martha had a half-dozen hair ribbons. She lost one. How many hair ribbons does she have left?

Number sentence ______________________

Answer ______________________

2. Draw pictures to show the number of jelly beans each child has.

Child	Number of Jelly Beans
Bill	10
Mike	7
Tia	8

Jelly Beans

Bill	
Mike	
Tia	

(oval) = 2 jelly beans

3. Find the differences.

$$\begin{array}{r} 51¢ \\ -\ 16¢ \\ \hline \end{array} \qquad \begin{array}{r} 70¢ \\ -\ 52¢ \\ \hline \end{array} \qquad \begin{array}{r} 67¢ \\ -\ 43¢ \\ \hline \end{array}$$

4. Put these numbers in order from least to greatest.
Circle the number that is between 27 and 35.

38	24	33	18	26

____ least ____ ____ ____ ____ greatest

5. Match the numbers with the words.

249 • • four hundred twenty-nine

429 • • two hundred ninety-four

492 • • two hundred forty-nine

294 • • four hundred ninety-two

6. Evan has 5 dimes and 14 pennies.
How much money is this?
Write the amount two ways. ____________ ____________

Name ______________________________

Class Fact Practice 92A

Saxon *Math 2 (for use with* ***Lesson 92****)*

Set 18: Subtracting 6 and 5

Do –6 Wrap-Up twice. ☐☐

Do –5 Wrap-Up twice. ☐☐

11 – 6 = ____	7 – 5 = ____
6 – 6 = ____	10 – 5 = ____
15 – 6 = ____	6 – 5 = ____
12 – 6 = ____	13 – 5 = ____
7 – 6 = ____	8 – 5 = ____
9 – 6 = ____	11 – 5 = ____
14 – 6 = ____	9 – 5 = ____
13 – 6 = ____	12 – 5 = ____
8 – 6 = ____	5 – 5 = ____
10 – 6 = ____	14 – 5 = ____

Name ______________________________

Saxon Math 2 (for use with **Lesson 92**)

Set 18: Subtracting 6 and 5

1. Write the answers.
2. Draw lines to connect the problems that have the same answer.
3. Ask someone to correct your paper. Corrected by ______________

$12 - 6 =$ ____ •	• $14 - 5 =$ ____
$6 - 6 =$ ____ •	• $7 - 5 =$ ____
$15 - 6 =$ ____ •	• $12 - 5 =$ ____
$8 - 6 =$ ____ •	• $11 - 5 =$ ____
$13 - 6 =$ ____ •	• $5 - 5 =$ ____
$10 - 6 =$ ____ •	• $6 - 5 =$ ____
$7 - 6 =$ ____ •	• $13 - 5 =$ ____
$14 - 6 =$ ____ •	• $9 - 5 =$ ____
$11 - 6 =$ ____ •	• $8 - 5 =$ ____
$9 - 6 =$ ____ •	• $10 - 5 =$ ____

Name. Draw a 3" line segment.

Date. Draw a $2\frac{1}{2}$" line segment.

Guided Class Practice 92A

***Saxon** Math 2 (for use with **Lesson 92**)*

1. Marsha counted the tiles and put them in groups of 10. When she finished, she counted 3 groups of tiles. Draw a picture to show the tiles.

Marsha has _____ groups of _____ tiles. How many tiles does she have? _______

2. Put these numbers in order from least to greatest.
Circle the number that is between 25 and 45.

71	47	24	78	38

_____ least _____ _____ _____ _____ greatest

3. Find the products.

$9 \times 10 =$ _______ $7 \times 10 =$ _______ $3 \times 10 =$ _______

4. Find the differences.

$$\begin{array}{r} 79¢ \\ -\ 34¢ \\ \hline \end{array} \qquad \begin{array}{r} 51¢ \\ -\ 13¢ \\ \hline \end{array} \qquad \begin{array}{r} 80¢ \\ -\ 24¢ \\ \hline \end{array}$$

5. Write the correct comparison symbol (>, <, or =).

21 ◯ 13   5 + 3 ◯ 10 − 2   1 × 10 ◯ 9 + 2

_____ _____   _____ _____

6. Find the sums.

$$\begin{array}{r} 23¢ \\ 42¢ \\ +\ 15¢ \\ \hline \end{array} \qquad \begin{array}{r} 49¢ \\ +\ 33¢ \\ \hline \end{array} \qquad \begin{array}{r} 37¢ \\ +\ 13¢ \\ \hline \end{array}$$

Name ______________________________

Date ______________________________

Homework 92B

***Saxon** Math 2 (for use with **Lesson 92**)*

1. Stephen counted the cubes and put them in groups of 10. When he finished, he counted 5 groups of cubes. Draw a picture to show the cubes.

Stephen has _____ groups of _____ cubes. How many cubes does he have? _______

2. Put these numbers in order from least to greatest.
Circle the number that is between 38 and 52.

37	26	54	29	47

_____ _____ _____ _____ _____
least ... greatest

3. Find the products.

$8 \times 10 =$ _______ $2 \times 10 =$ _______ $4 \times 10 =$ _______

4. Find the differences.

84¢ − 55¢ = ____ 56¢ − 22¢ = ____ 60¢ − 13¢ = ____

5. Write the correct comparison symbol (>, <, or =).

14 ◯ 25 $3 + 6$ ◯ $10 - 3$ $7 + 5$ ◯ $4 + 8$

6. Find the sums.

65¢ + 73¢ = ____ 38¢ + 57¢ = ____ 23¢ + 48¢ + 17¢ = ____

Name ______________________________

Class Fact Practice 93A

***Saxon** Math 2 (for use with **Lesson 93**)*

Set 18: Subtracting 6 and 5

$6 - 6$	$11 - 6$	$12 - 5$	$15 - 6$	$8 - 5$
$13 - 6$	$10 - 6$	$6 - 5$	$12 - 6$	$9 - 5$
$7 - 6$	$14 - 6$	$11 - 5$	$9 - 6$	$13 - 5$
$15 - 6$	$10 - 5$	$14 - 5$	$11 - 6$	$14 - 6$
$7 - 5$	$12 - 6$	$8 - 6$	$5 - 5$	$13 - 6$

Name ____________________

Fact Homework 93B

Saxon Math 2 (for use with **Lesson 93**)

Set 18: Subtracting 6 and 5

1. Read the answers to someone.
2. Write the answers.
3. Ask someone to correct your paper. Corrected by ____________________

13 − 6	5 − 5	8 − 6	12 − 6	7 − 5
14 − 6	11 − 6	14 − 5	10 − 5	15 − 6
13 − 5	9 − 6	11 − 5	14 − 6	7 − 6
9 − 5	12 − 6	6 − 5	10 − 6	13 − 6
8 − 5	15 − 6	12 − 5	11 − 6	6 − 6

Name

Draw a line segment 1" shorter than the date line segment.

Date

1. Ellen has 5 dimes and 8 pennies. Kay has 2 dimes and 14 pennies. How much money does each girl have?

Ellen ________ Kay ________

Who has the most money? ________

2. How much money is this? ________

3. What fractional part of the circles is colored? ___

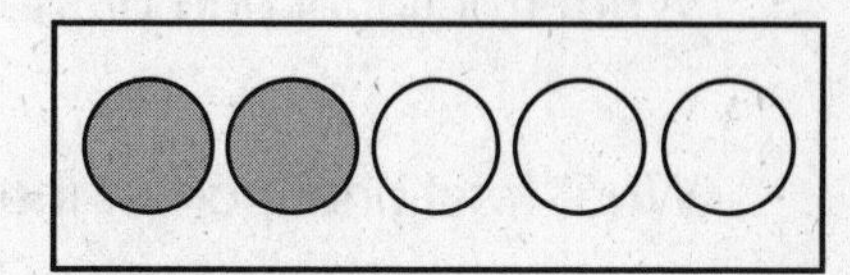

What fractional part of the circles is not colored? ___

4. Draw a picture to show 416. (Use ☐ for 100, ▯ for 10, and ▫ for 1.)

Write 416 in expanded form. ________

5. Circle the shape which shows a line of symmetry.

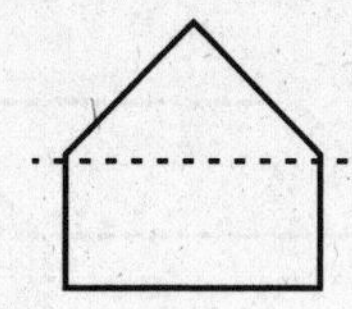

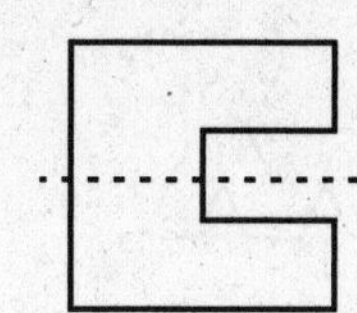

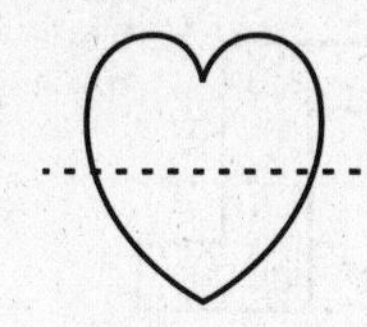

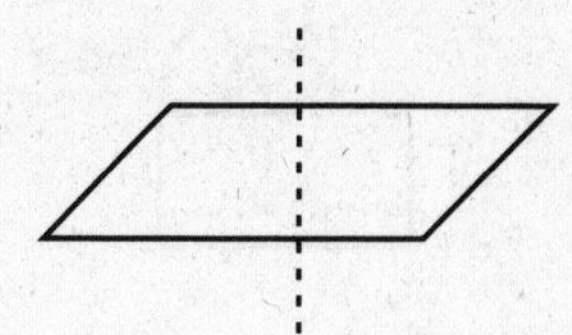

6. Find the answers.

49¢	70¢	51¢	65¢
− 25¢	− 23¢	− 36¢	29¢
			+ 51¢

Name ______________________

Date ______________________

1. Andrew has 2 dimes and 17 pennies. Danny has 3 dimes and 5 pennies. How much money does each boy have?

Andrew __________ Danny __________

Who has the most money? __________

2. How much money is this? __________

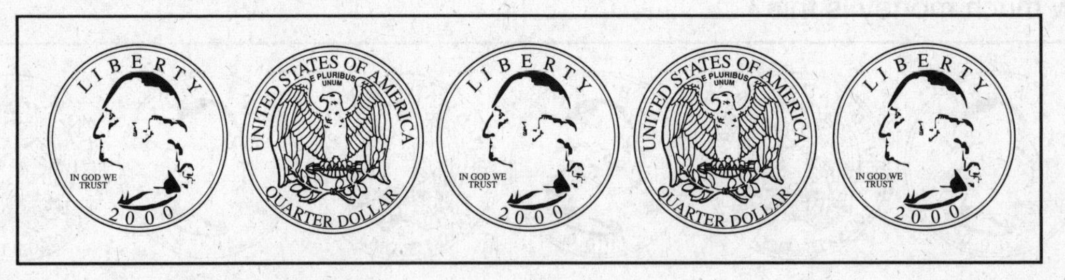

3. What fractional part of the circles is colored? ____

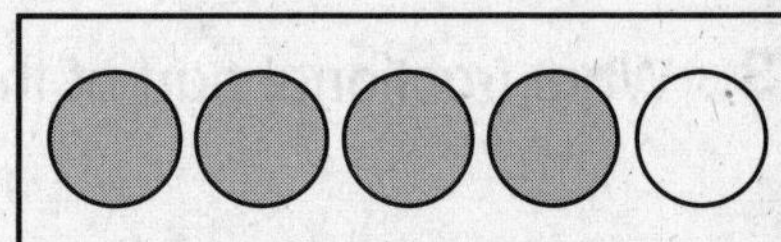

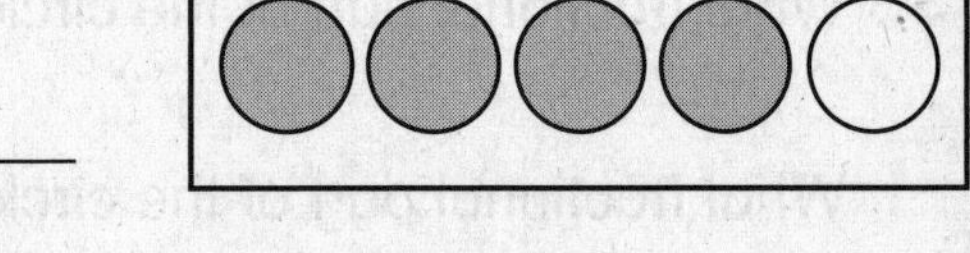

What fractional part of the circles is not colored? ____

4. Draw a picture to show 203. (Use ☐ for 100, ▯ for 10, and ▫ for 1.)

Write 203 in expanded form. __________

5. Circle the shape which shows a line of symmetry.

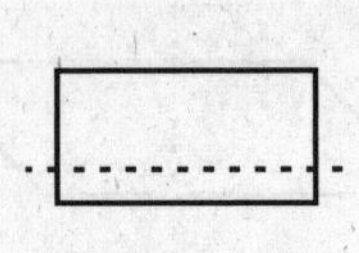 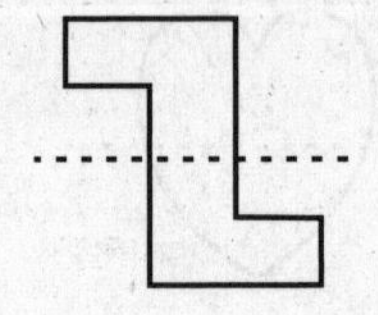 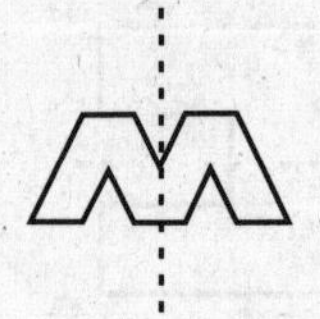 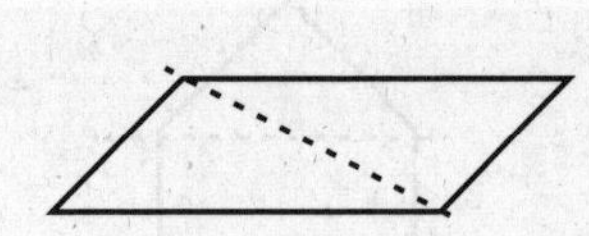

6. Find the answers.

42¢	40¢	65¢	74¢
− 15¢	− 23¢	− 43¢	23¢
			+ 34¢

Name ______________________________

Class Fact Practice 94A

***Saxon** Math 2 (for use with **Lesson 94**)*

Set 18: Subtracting 6 and 5

Do –6 Wrap-Up twice.
Do –5 Wrap-Up once.

$8 - 6$	$10 - 5$	$12 - 6$	$8 - 5$	$6 - 6$
$15 - 6$	$12 - 5$	$7 - 6$	$14 - 5$	$11 - 6$
$7 - 5$	$13 - 6$	$9 - 5$	$9 - 6$	$6 - 5$
$10 - 6$	$13 - 5$	$14 - 6$	$11 - 5$	$5 - 5$
$14 - 5$	$12 - 6$	$7 - 5$	$14 - 6$	$9 - 6$

Name ______________________________

Saxon Math 2 (for use with **Lesson 94**)

Set 18: Subtracting 6 and 5

1. Read the answers to someone.
2. Write the answers.
3. Ask someone to correct your paper. Corrected by ______________________

$9 - 5$	$7 - 6$	$5 - 5$	$14 - 5$	$11 - 6$
$12 - 5$	$13 - 6$	$10 - 5$	$8 - 6$	$6 - 5$
$9 - 6$	$13 - 5$	$12 - 6$	$7 - 5$	$10 - 6$
$15 - 6$	$11 - 5$	$6 - 6$	$14 - 6$	$8 - 5$
$14 - 6$	$7 - 5$	$9 - 6$	$12 - 6$	$13 - 5$

Name . Draw a $2\frac{1}{2}$" line segment.

Guided Class Practice 94A

***Saxon** Math 2 (for use with **Lesson 94**)*

Date . Draw a line segment 1" longer than the line segment for your name.

1. The cost of a small notebook is 67¢. The cost of a pencil is 23¢. How much money does Stephen need to buy a small notebook and a pencil?

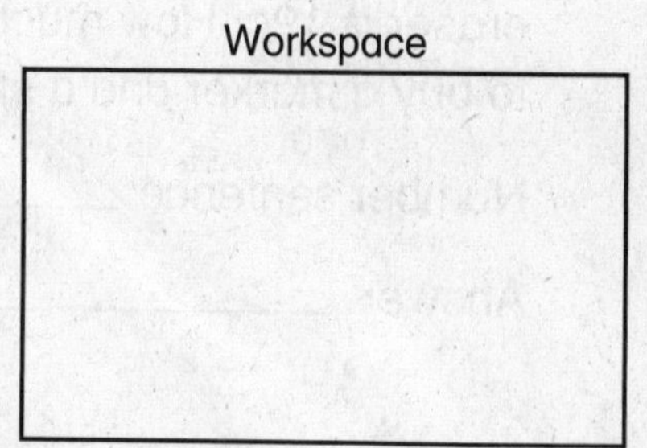

Number sentence ______________

Answer ______________

2. Each number is between what two tens? Circle the 10 each number is closest to.

______, 17, ______ ______, 43, ______ ______, 36, ______

3. The children in Mrs. McCluckie's class counted twelve blue cars, eight red cars, and five gray cars in the school parking lot. Draw pictures on the graph to show how many cars they counted.

Colors of Cars

Red	
Blue	
Gray	

= 2 cars

4. Write the correct comparison symbol (>, <, or =).

15 – 6 ◯ 3 + 4 47 – 10 ◯ 39 + 1

______ ______ ______ ______

5. Circle the tool you would use to measure the temperature.

ruler measuring cup balance thermometer

6. Find the answers.

31 – 14

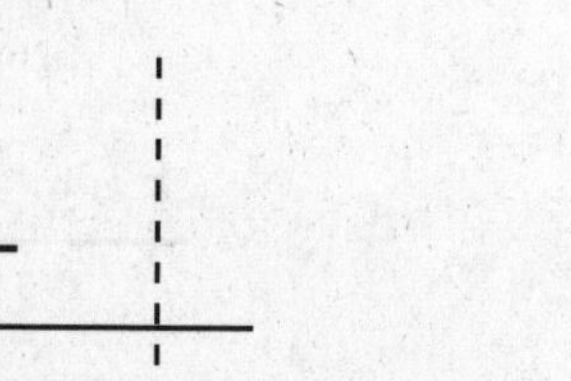

74 – 43

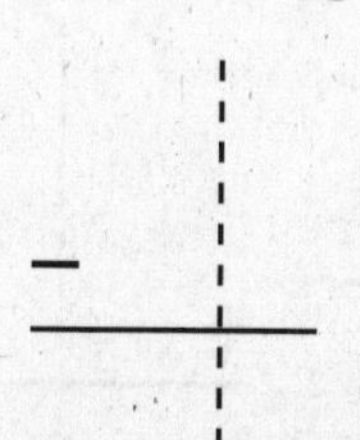

52 + 39 + 48

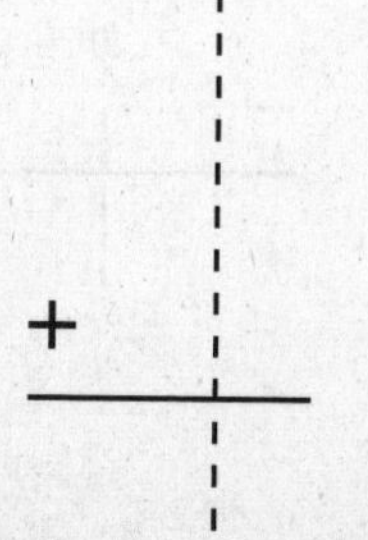

7 + 36

+

Name ______________________________

Date ______________________________

Saxon Math 2 (for use with **Lesson 94**)

1. The cost of a marker is 74¢. The cost of a small eraser is 19¢. How much money does DeAnna need to buy a marker and a small eraser?

Workspace

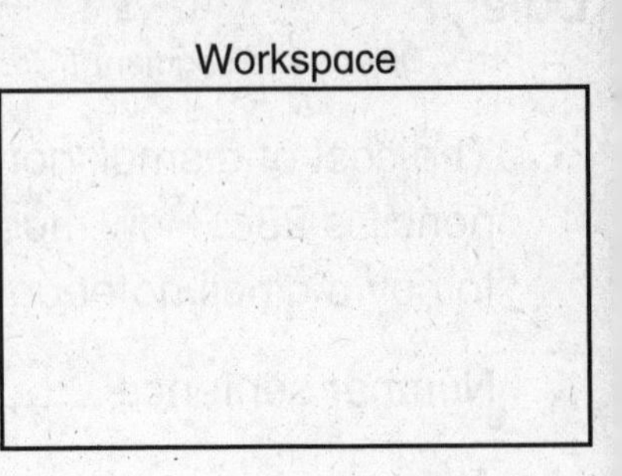

Number sentence ______________________

Answer ______________________

2. Each number is between what two tens? Circle the 10 each number is closest to.

_____, 24, _____ _____, 58, _____ _____, 27, _____

3. The children in Mrs. Casner's class counted six blue cars, four red cars, and three gray cars. Draw pictures on the graph to show how many cars they counted.

Colors of Cars

Red	
Blue	
Gray	

= 2 cars

4. Write the correct comparison symbol (>, <, or =).

13 – 6 ◯ 9 – 3 29 – 10 ◯ 18 + 1

5. Circle the tool you would use to measure how much water is in a bottle?

ruler measuring cup balance thermometer

6. Find the answers.

43 – 29 82 – 51 31 + 49 + 24 6 + 58

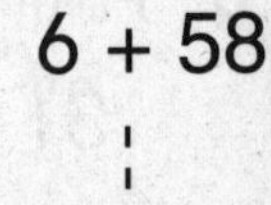

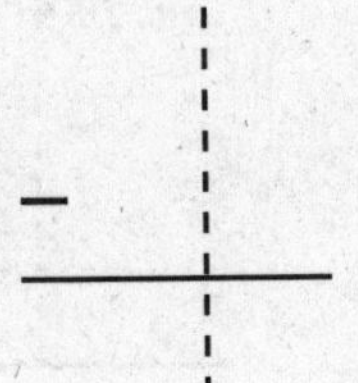

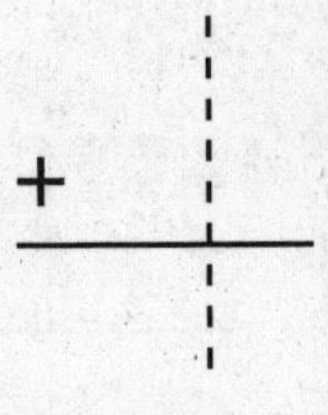

Name ______________________ Score ________

Fact Assessment 18-1

Saxon *Math 2 (for use with* ***Lesson 95-1****)*

Set 18: Subtracting 6 and 5

$7 - 6$	$13 - 6$	$10 - 6$	$6 - 5$	$8 - 6$
$15 - 6$	$9 - 5$	$6 - 6$	$11 - 6$	$7 - 5$
$9 - 6$	$14 - 5$	$12 - 6$	$7 - 6$	$10 - 5$
$14 - 6$	$12 - 5$	$6 - 6$	$11 - 5$	$8 - 6$
$10 - 6$	$8 - 5$	$9 - 6$	$13 - 5$	$5 - 5$

Name ______________________________

Fact Homework 95A

Saxon *Math 2 (for use with* ***Lesson 95-1****)*

Set 18: Subtracting 6 and 5

1. Pretend you are the teacher.
2. Correct this paper.
3. If the answer is incorrect, write the correct answer next to the problem.

7 − 6 = 1	13 − 6 = 8	10 − 6 = 4	6 − 5 = 1	8 − 6 = 2
15 − 6 = 9	9 − 5 = 4	6 − 6 = 0	11 − 6 = 4	7 − 5 = 2
9 − 6 = 3	14 − 5 = 8	12 − 6 = 6	7 − 6 = 1	10 − 5 = 5
14 − 6 = 9	12 − 5 = 7	6 − 6 = 0	11 − 5 = 6	8 − 6 = 3
10 − 6 = 4	8 − 5 = 3	9 − 6 = 3	13 − 5 = 8	5 − 5 = 0

Name ____________________

Fact Homework 95B

Saxon Math 2 (for use with Lesson 95-1)

S70: 70 Subtraction Facts   Corrected by ________

Row										
1	7 − 1	10 − 4	9 − 0	5 − 4	12 − 6	11 − 3	8 − 2	6 − 6	9 − 4	6 − 3
2	11 − 6	10 − 2	6 − 1	2 − 0	9 − 3	7 − 2	6 − 5	5 − 5	11 − 4	4 − 2
3	8 − 0	10 − 6	14 − 5	4 − 0	9 − 5	8 − 4	9 − 2	11 − 5	15 − 6	5 − 1
4	8 − 5	8 − 6	1 − 0	7 − 3	9 − 6	4 − 3	10 − 5	12 − 4	8 − 3	10 − 3
5	4 − 1	6 − 2	13 − 5	7 − 0	11 − 2	14 − 6	1 − 1	12 − 3	7 − 5	2 − 1
6	2 − 2	12 − 5	3 − 1	10 − 1	6 − 0	13 − 4	5 − 2	3 − 0	7 − 6	13 − 6
7	3 − 3	9 − 1	6 − 4	7 − 4	4 − 4	3 − 2	5 − 0	5 − 3	0 − 0	8 − 1

Name ______________________

Lesson Worksheet 95-1A

***Saxon** Math 2 (for use with **Lesson 95-1**)*

A. Write the answers.

7 – 7 = ______

8 – 7 = ______

9 – 7 = ______

10 – 7 = ______

11 – 7 = ______

12 – 7 = ______

13 – 7 = ______

14 – 7 = ______

15 – 7 = ______

16 – 7 = ______

17 – 7 = ______

18 – 7 = ______

B. Draw lines to connect the problems to the answers.

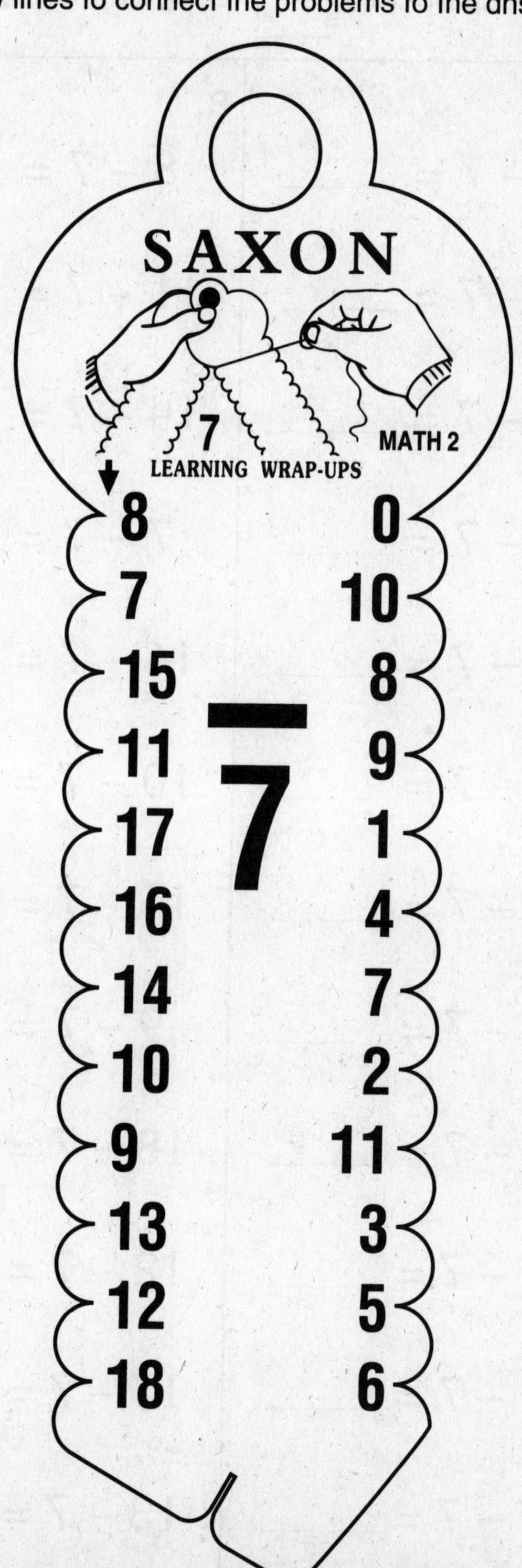

Name ______________________

Lesson Worksheet 95-1

Saxon *Math 2 (for use with* ***Lesson 95-1****)*

Do –7 Wrap-Up once. ☐ Do –7 Wrap-Up once. ☐ Do –7 Wrap-Up once. ☐

A.	B.	C.
10 – 7 = ____	8 – 7 = ____	12 – 7 = ____
15 – 7 = ____	11 – 7 = ____	7 – 7 = ____
9 – 7 = ____	14 – 7 = ____	15 – 7 = ____
12 – 7 = ____	7 – 7 = ____	10 – 7 = ____
18 – 7 = ____	16 – 7 = ____	17 – 7 = ____
11 – 7 = ____	10 – 7 = ____	8 – 7 = ____
13 – 7 = ____	15 – 7 = ____	16 – 7 = ____
7 – 7 = ____	9 – 7 = ____	13 – 7 = ____
14 – 7 = ____	18 – 7 = ____	9 – 7 = ____
16 – 7 = ____	12 – 7 = ____	14 – 7 = ____
8 – 7 = ____	17 – 7 = ____	18 – 7 = ____
17 – 7 = ____	13 – 7 = ____	11 – 7 = ____

Name ______________________

Measure this line segment using inches. ____"

Date ______________

Measure this line segment using inches. ____"

1. Erica went to a party at 4:00 p.m. She left when the party was over at 7:00 p.m.

How long was she at the party? ________________

2. Write 437 in expanded form. ______________________

Write the number for 200 + 60 + 7. __________

3. Draw a pictograph to show how many children like each ice cream flavor.

Ice Cream Flavor	Number of Children
Vanilla	8
Chocolate	10
Strawberry	7

	Children
Vanilla	
Chocolate	
Strawberry	

☺ = 2 children

4. Match the numbers with the words.

607 •	• six hundred seventy
706 •	• seven hundred sixty
760 •	• six hundred seven
670 •	• seven hundred six

5. My favorite time of day is 8:35 p.m.
Show that time on the clock.

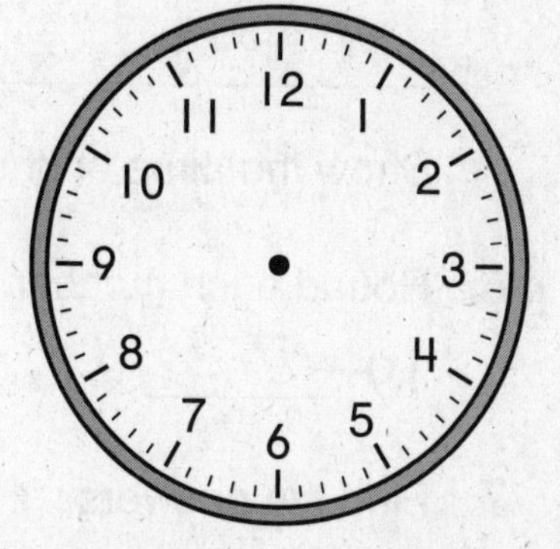

Is it morning or evening? ________________

6. Round each number to the nearest 10.

27 ______ 52 ______ 45 ______

7. Find the answers.

58	47¢	9¢	80¢
+ 16	− 16¢	+ 42¢	− 35¢

Name ______________________________

Date ______________________________

Homework 95B

***Saxon** Math 2 (for use with **Lesson 95-1**)*

1. Luis went to visit his grandfather at 3:00 p.m. He left for home at 8:00 p.m.

How long was he at his grandfather's? ______________

2. Write 420 in expanded form. ______________________________

Write the number for 6 + 300 + 40. __________

3. Draw a pictograph to show how many children like each ice cream flavor.

Ice Cream Flavor	Number of Children
Chocolate Chip	12
Mint Chip	6
Butter Crunch	5

	Children
Chocolate Chip	
Mint Chip	
Butter Crunch	

☺ = 2 children

4. Match the numbers with the words.

901 •	• one hundred nine
910 •	• one hundred ninety
190 •	• nine hundred ten
109 •	• nine hundred one

5. What is your favorite time of the day?

____

Show that time on the clock.

6. Round each number to the nearest 10.

19 ________ 63 ________ 75 ________

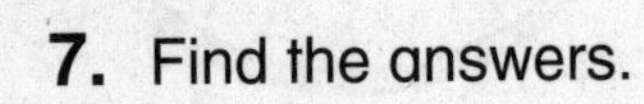

7. Find the answers.

$$\begin{array}{r} 23 \\ +\ 48 \\ \hline \end{array} \qquad \begin{array}{r} 28¢ \\ -\ 15¢ \\ \hline \end{array} \qquad \begin{array}{r} 8¢ \\ +\ 29¢ \\ \hline \end{array} \qquad \begin{array}{r} 70¢ \\ -\ 34¢ \\ \hline \end{array}$$

Name ______________________ Score ______

Fact Assessment 18-2

***Saxon** Math 2 (for use with **Lesson 95-2**)*

S70: 70 Subtraction Facts

1	$7 - 1$	$10 - 4$	$9 - 0$	$5 - 4$	$12 - 6$	$11 - 3$	$8 - 2$	$6 - 6$	$7 - 4$	$6 - 3$
2	$11 - 6$	$10 - 2$	$6 - 1$	$2 - 0$	$9 - 3$	$7 - 2$	$6 - 5$	$5 - 5$	$11 - 4$	$4 - 2$
3	$8 - 0$	$10 - 6$	$14 - 5$	$4 - 0$	$9 - 5$	$8 - 4$	$9 - 2$	$11 - 5$	$15 - 6$	$5 - 1$
4	$8 - 5$	$8 - 6$	$1 - 0$	$7 - 3$	$9 - 6$	$4 - 3$	$10 - 5$	$12 - 4$	$8 - 3$	$10 - 3$
5	$4 - 1$	$6 - 2$	$13 - 5$	$7 - 0$	$11 - 2$	$14 - 6$	$1 - 1$	$12 - 3$	$7 - 5$	$2 - 1$
6	$2 - 2$	$12 - 5$	$3 - 1$	$10 - 1$	$6 - 0$	$13 - 4$	$5 - 2$	$3 - 0$	$7 - 6$	$13 - 6$
7	$3 - 3$	$9 - 1$	$6 - 4$	$9 - 4$	$4 - 4$	$3 - 2$	$5 - 0$	$5 - 3$	$0 - 0$	$8 - 1$

Name ______________________

Date ______________________

1. There were 42 children on the bus. Ten children got off the bus. How many children are on the bus now?

Number sentence ______________________

Answer ______________________

2. Draw a pictograph to show how many cookies each child ate.

Child	Number of Cookies
Jacob	10
Ashley	6
Fernando	7

Number of Cookies Eaten

Jacob	
Ashley	
Fernando	

○ = 2 cookies

3. Use the correct comparison symbol (>, <, or =).

16 ○ 9   3 + 7 ○ 10   8 + 1 ○ 8 + 2

4. I have 2 dimes, 3 nickels, and 4 pennies. Draw the coins. How much money is this?

Write the amount two ways. __________ __________

5. Draw a picture to show 273. (Use □ for 100, ▯ for 10, and ▪ for 1.)

Write 273 in expanded form. ______________________

6. Find the sums.

$$\begin{array}{r} 35¢ \\ +\ 87¢ \\ \hline \end{array} \qquad \begin{array}{r} 59¢ \\ +\ 41¢ \\ \hline \end{array} \qquad \begin{array}{r} 22¢ \\ 6¢ \\ +\ 53¢ \\ \hline \end{array} \qquad \begin{array}{r} 62¢ \\ 37¢ \\ +\ 43¢ \\ \hline \end{array}$$

Name ______________________________

Lesson Worksheet 95-2

***Saxon** Math 2 (for use with **Lesson 95-2**)*

Bag	Estimate	Actual
A		
B		
C		
D		
E		
F		
G		

Picture for Bag ________

Name ____________________

Class Fact Practice 96A

***Saxon** Math 2 (for use with **Lesson 96**)*

Set 19: Subtracting 7 and 6; Review Facts

Do −7 Wrap-Up once.
Do −6 Wrap-Up once.

$8 - 7$	$11 - 7$	$16 - 7$	$13 - 7$	$10 - 7$
$15 - 7$	$9 - 7$	$14 - 7$	$7 - 7$	$12 - 7$
$12 - 6$	$7 - 6$	$15 - 6$	$10 - 6$	$6 - 6$
$13 - 6$	$9 - 6$	$11 - 6$	$8 - 6$	$14 - 6$
$14 - 5$	$8 - 5$	$5 - 5$	$10 - 5$	$12 - 5$

Name ______________________________

Set 19: Subtracting 7 and 6; Review Facts

1. Read the answers to someone.
2. Write the answers.
3. Ask someone to correct your paper. Corrected by ____________________

13 − 7	10 − 7	16 − 7	8 − 7	12 − 7
14 − 7	7 − 7	15 − 7	11 − 7	9 − 7
15 − 6	6 − 6	11 − 6	14 − 6	9 − 6
12 − 6	8 − 6	13 − 6	7 − 6	10 − 6
10 − 5	8 − 5	6 − 5	13 − 5	9 − 5

Name ______________________________

Lesson Worksheet 96

***Saxon** Math 2 (first used in **Lesson 96**)*

one half of ______ is ______

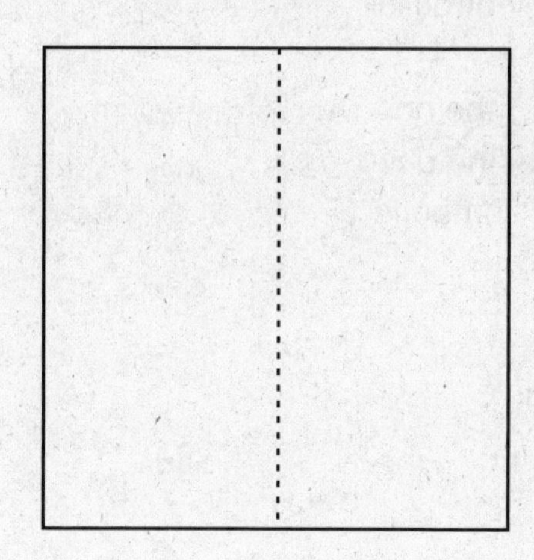

one half of ______ is ______

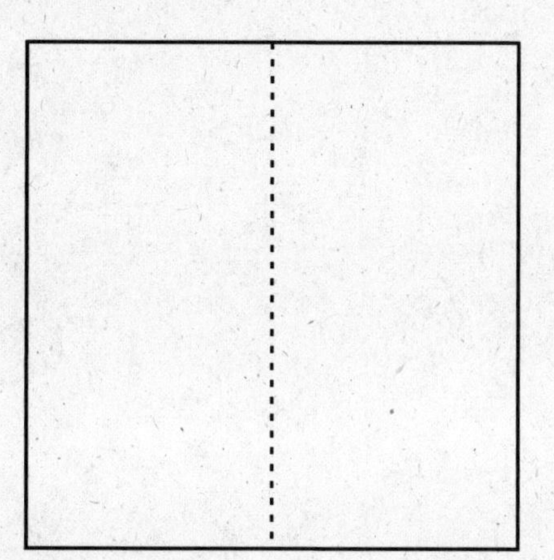

one half of ______ is ______

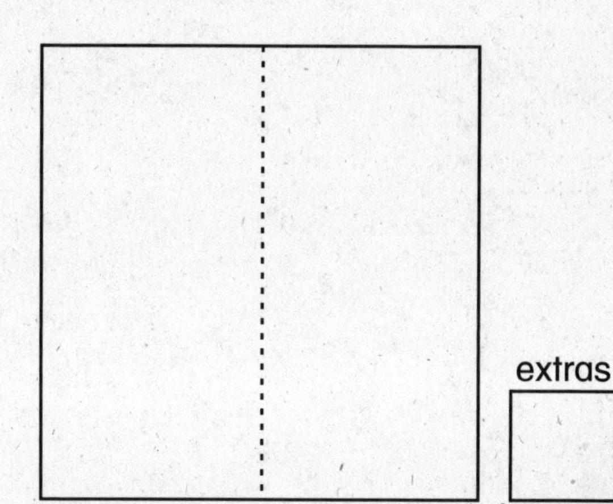

extras

one half of ______ is ______

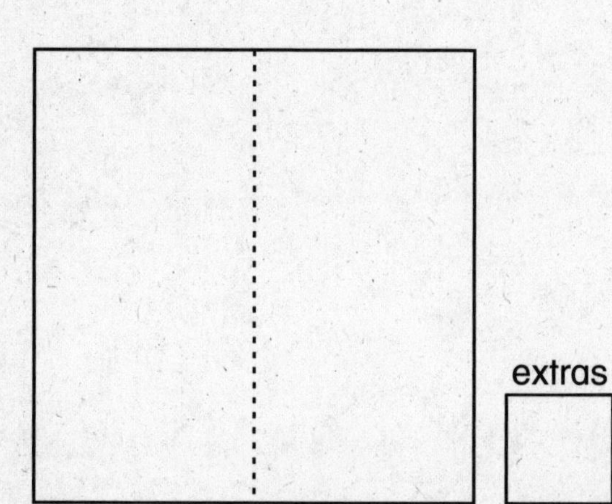

extras

one half of ______ is ______

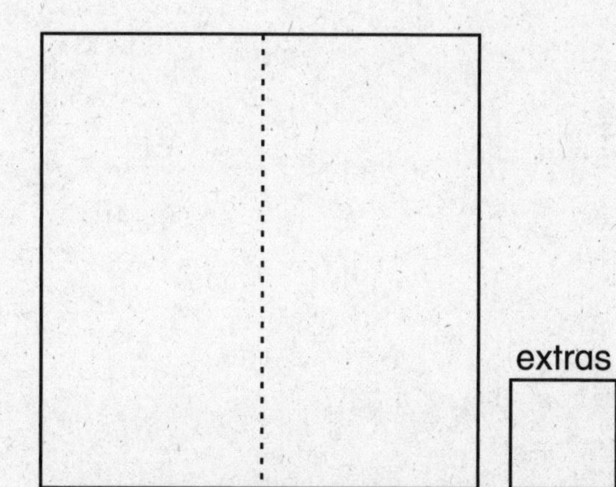

extras

one half of ______ is ______

Name ____________________

Measure this line segment using inches. ________"

Date ____________________

Measure this line segment using inches. ________"

1. Dolores has 3 dimes and 6 pennies. Frances has 7 pennies and 2 dimes. How much money does each girl have?

Dolores __________ Frances __________

How much money do they have altogether?

Number sentence ____________________

Answer ____________________

Workspace

2. Put these numbers in order from least to greatest. Circle the number that is between 250 and 280.

176	284	373	181	279

_____ _____ _____ _____ _____
least greatest

3. Show how two children will share ten books equally.

☐ = 1 book

How many books will each child have? ______

one half of 10 is ______

4. Round each number to the nearest 10.

47 ______ 12 ______ 85 ______

5. Draw a dozen eggs.

Color five eggs red.

What fractional part of the eggs is red? ____

6. Find the answers.

$$\begin{array}{r} 74 \\ -\ 47 \\ \hline \end{array} \qquad \begin{array}{r} 60 \\ -\ 47 \\ \hline \end{array} \qquad \begin{array}{r} 96 \\ 25 \\ +\ 32 \\ \hline \end{array}$$

$6 + 9 + 2 + 5 + 4 + 1 =$ ______

Name ______________________________

Date ______________________________

Homework 96B

***Saxon** Math 2 (for use with **Lesson 96**)*

1. Juan has 6 pennies and 4 dimes. Gary has 6 dimes and 9 pennies. How much money does each boy have?

Workspace

Juan __________ Gary __________

How much money do they have altogether?

Number sentence ______________________

Answer ______________________

2. Put these numbers in order from least to greatest. Circle the number that is between 300 and 500.

192	284	194	487	291

______ ______ ______ ______ ______

least greatest

3. Show how two children will share six books equally.

How many books will each child have? __________

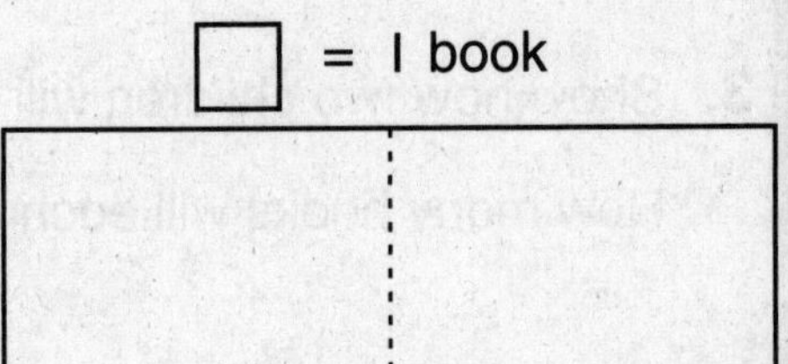

one half of 6 is ______

4. Round each number to the nearest 10.

66 ______ 8 ______ 75 ______

5. Draw a half-dozen eggs.

Color five eggs red.

What fractional part of the eggs is red? ______

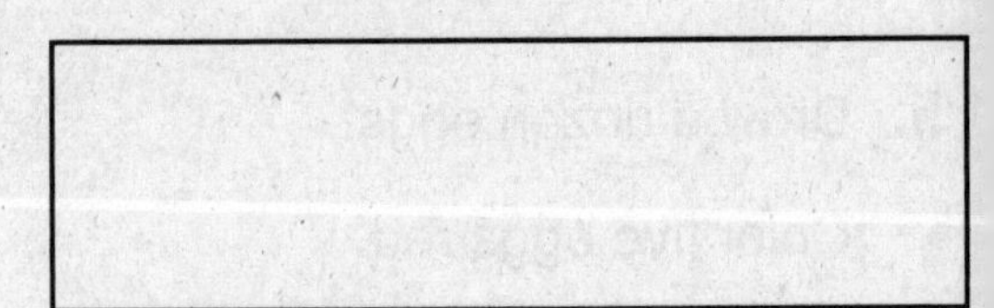

6. Find the answers.

$$\begin{array}{r} 51 \\ -\ 17 \\ \hline \end{array} \qquad \begin{array}{r} 73 \\ -\ 37 \\ \hline \end{array} \qquad \begin{array}{r} 72 \\ 24 \\ +\ 47 \\ \hline \end{array}$$

4 + 8 + 2 + 7 + 3 + 1 = ______

Name ______________________________

Class Fact Practice 97A

***Saxon** Math 2 (for use with **Lesson 97**)*

Set 19: Subtracting 7 and 6

Do –7 Wrap-Up twice. ☐☐ Do –6 Wrap-Up twice. ☐☐

9 – 7 = ____	14 – 6 = ____
12 – 7 = ____	8 – 6 = ____
7 – 7 = ____	11 – 6 = ____
16 – 7 = ____	6 – 6 = ____
11 – 7 = ____	10 – 6 = ____
8 – 7 = ____	13 – 6 = ____
14 – 7 = ____	7 – 6 = ____
10 – 7 = ____	12 – 6 = ____
15 – 7 = ____	9 – 6 = ____
13 – 7 = ____	15 – 6 = ____

Name ______________________________

Fact Homework 97B

Saxon *Math 2 (for use with* ***Lesson 97****)*

Set 19: Subtracting 7 and 6

1. Write the answers.
2. Draw lines to connect the problems that have the same answer.
3. Ask someone to correct your paper. Corrected by ____________________

7 – 7 = ____ •	• 13 – 6 = ____
11 – 7 = ____ •	• 8 – 6 = ____
14 – 7 = ____ •	• 6 – 6 = ____
9 – 7 = ____ •	• 10 – 6 = ____
16 – 7 = ____ •	• 14 – 6 = ____
12 – 7 = ____ •	• 15 – 6 = ____
8 – 7 = ____ •	• 7 – 6 = ____
15 – 7 = ____ •	• 11 – 6 = ____
10 – 7 = ____ •	• 12 – 6 = ____
13 – 7 = ____ •	• 9 – 6 = ____

Name
Draw a $3\frac{1}{2}$" line segment.

Date
Measure this line segment using inches. ______"

1. There were a dozen children in the pool. A half-dozen children left the pool. How many children are in the pool now?

Number sentence ______________________

Answer ______________

2. Color the thermometer to show 84°F.

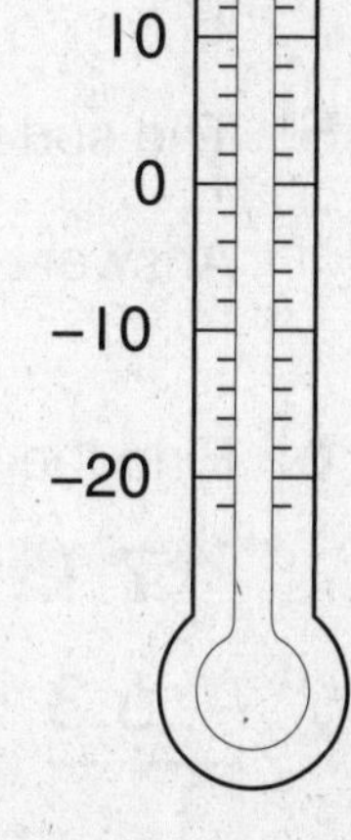

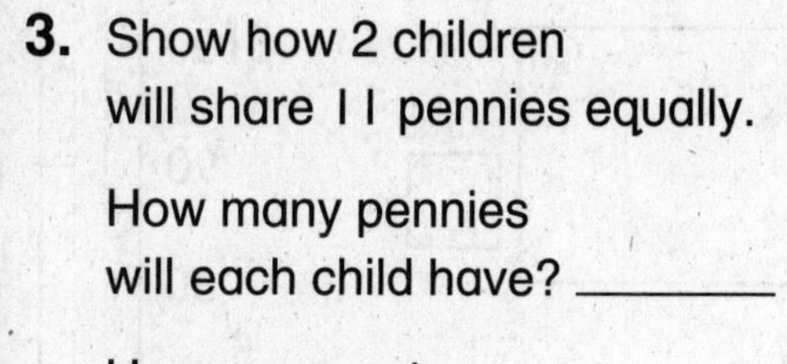

3. Show how 2 children will share 11 pennies equally.

How many pennies will each child have? ______

How many extra pennies are there? ______

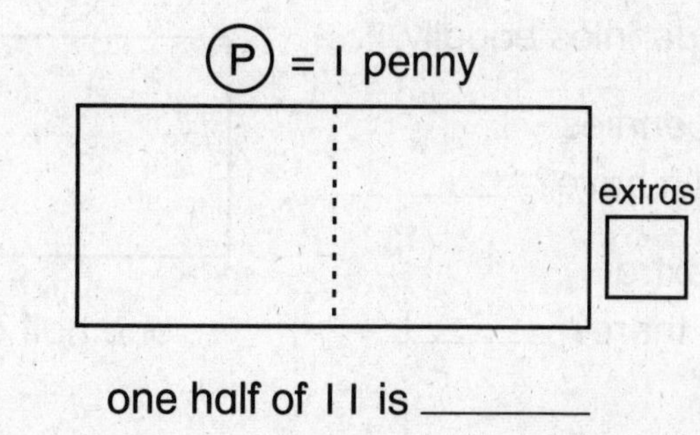

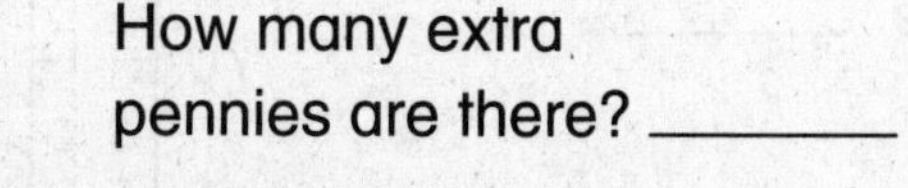

one half of 11 is ______

4. Write 806 in expanded form. ______________

What digit is in the hundreds' place? ______

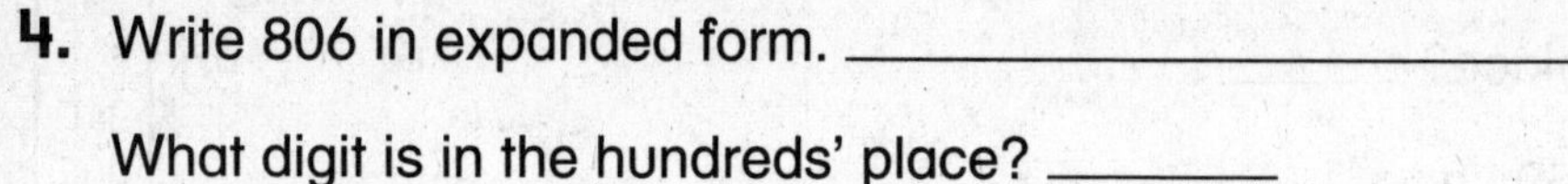

Draw a picture to show 806.

(Use ☐ for 100, ▯ for 10, and ▫ for 1.)

5. I can see the stars. What time is it?

Answer ______________

6. Find the differences.

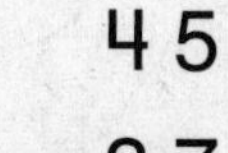

$$\begin{array}{r} 45 \\ -\ 37 \\ \hline \end{array} \qquad \begin{array}{r} 71 \\ -\ 53 \\ \hline \end{array} \qquad \begin{array}{r} 85 \\ -\ 32 \\ \hline \end{array}$$

Name ______________________

Date ______________________

Saxon Math 2 (for use with **Lesson 97**)

1. There were a half-dozen children in the gym. Another half-dozen children joined them. How many children are in the gym now?

Number sentence ______________________

Answer ______________________

2. Color the thermometer to show 68°F.

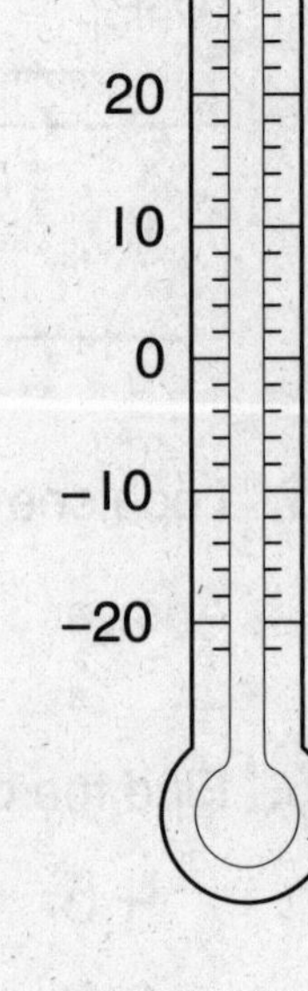

3. Show how 2 children will share 9 pennies equally.

How many pennies will each child have? ________

How many extra pennies are there? ________

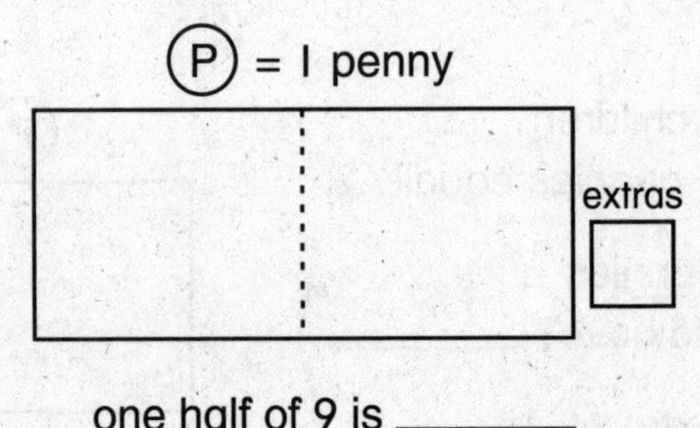

one half of 9 is ________

4. Write 350 in expanded form. ______________________

What digit is in the tens' place? ________

Draw a picture to show 350.

(Use [square] for 100, [bar] for 10, and ▫ for 1.)

5. The sun is shining. What time is it?

Answer ______________________

6. Find the differences.

$$\begin{array}{r} 71 \\ -\ 63 \\ \hline \end{array} \qquad \begin{array}{r} 44 \\ -\ 16 \\ \hline \end{array} \qquad \begin{array}{r} 58 \\ -\ 37 \\ \hline \end{array}$$

Name ______________________________

Class Fact Practice 98A

***Saxon** Math 2 (for use with **Lesson 98**)*

Set 19: Subtracting 7 and 6

12 − 7	7 − 6	15 − 7	7 − 7	9 − 6
14 − 7	12 − 6	8 − 7	10 − 6	13 − 7
11 − 6	16 − 7	13 − 6	9 − 7	15 − 6
13 − 7	10 − 7	8 − 6	14 − 7	12 − 7
6 − 6	15 − 7	11 − 7	14 − 6	16 − 7

Name ______________________________

Set 19: Subtracting 7 and 6

1. Read the answers to someone.
2. Write the answers.
3. Ask someone to correct your paper. Corrected by ____________________

9 − 6	7 − 7	15 − 7	7 − 6	12 − 7
13 − 7	10 − 6	8 − 7	12 − 6	14 − 7
15 − 6	9 − 7	13 − 6	16 − 7	11 − 6
12 − 7	14 − 7	8 − 6	10 − 7	13 − 7
16 − 7	14 − 6	11 − 7	15 − 7	6 − 6

Name •

Draw a 3" line segment.

Date •

Draw a line segment that is one inch shorter than the line segment for your name.

1. Amber has 5 dimes and 8 pennies. Eileen has 6 pennies and 3 dimes. How much money does each girl have?

Workspace

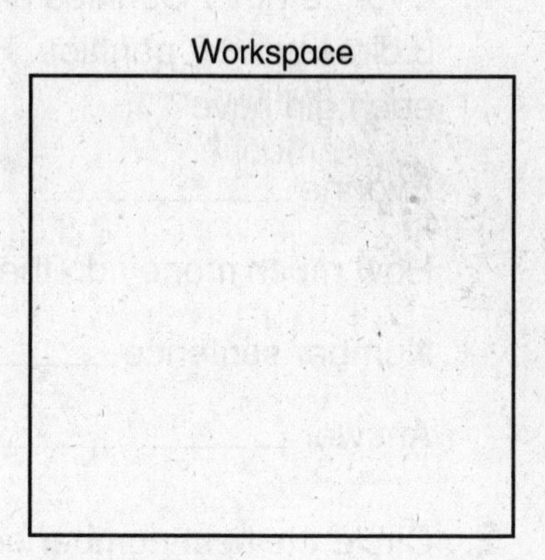

Amber ____________ Eileen ____________

How much money do they have altogether?

Number sentence ______________________________

Answer ______________________________

2. Circle the best number sentence to use to estimate the sum of 23 and 57.

$20 + 50 = 70$ $20 + 60 = 80$ $30 + 50 = 80$ $30 + 60 = 90$

3. Color $\frac{3}{4}$ of the circles.

4. Show how two children will share the markers equally.

9 markers

extras

one half of 9 is ______

14 markers

extras

one half of 14 is ______

5. Write this money amount in two ways. ____________ ____________

6. Find the answers.

$$\begin{array}{r} 48 \\ +\ 89 \\ \hline \end{array} \qquad \begin{array}{r} 93 \\ +\ 77 \\ \hline \end{array} \qquad \begin{array}{r} 91 \\ -\ 85 \\ \hline \end{array} \qquad \begin{array}{r} 67 \\ -\ 24 \\ \hline \end{array}$$

Name ____________________

Date ____________________

1. Evonne has 7 pennies and 4 dimes. Joann has 3 dimes and 9 pennies. How much money does each girl have?

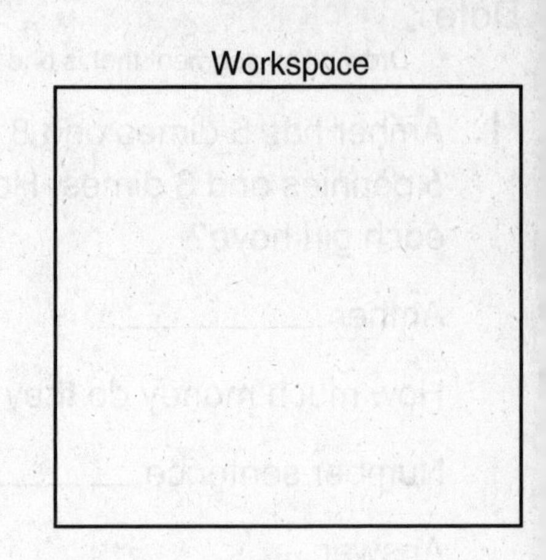

Evonne ____________ Joann ____________

How much money do they have altogether?

Number sentence ____________________

Answer ____________________

2. Circle the best number sentence to use to estimate the sum of 27 and 52.

$20 + 50 = 70$ $20 + 60 = 80$ $30 + 50 = 80$ $30 + 60 = 90$

3. Color $\frac{1}{4}$ of the circles.

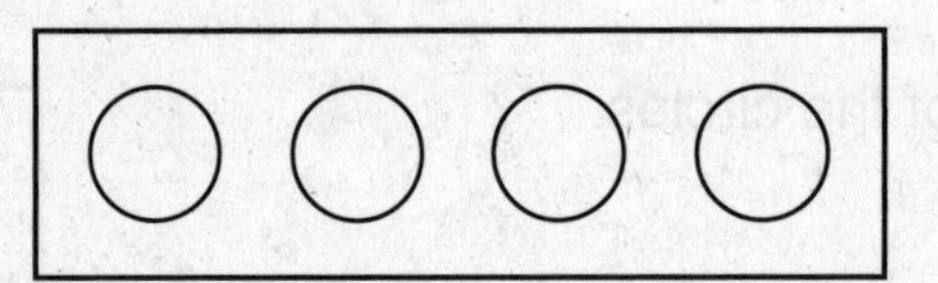

4. Show how two children will share the markers equally.

10 markers

extras

one half of 10 is ______

7 markers

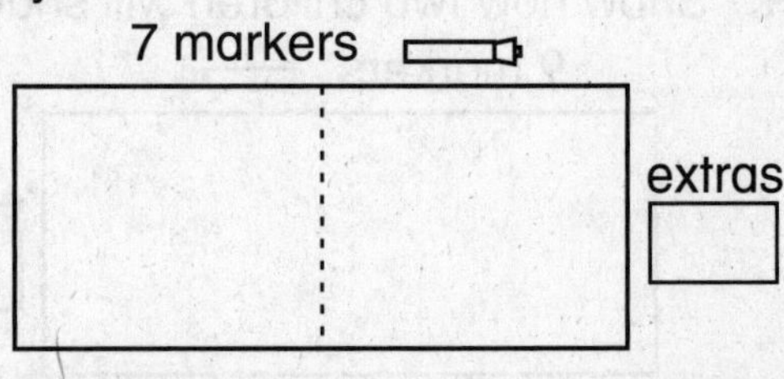

one half of 7 is ______

5. Write this money amount in two ways. ____________

6. Find the answers.

$\begin{array}{r} 64 \\ +\ 77 \\ \hline \end{array}$ $\begin{array}{r} 82 \\ +\ 97 \\ \hline \end{array}$ $\begin{array}{r} 62 \\ -\ 54 \\ \hline \end{array}$ $\begin{array}{r} 59 \\ -\ 27 \\ \hline \end{array}$

Name ______________________________

Class Fact Practice **99A**

Saxon *Math 2 (for use with* ***Lesson 99****)*

Set 19: Subtracting 7 and 6

Do –7 Wrap-Up twice.
Do –6 Wrap-Up once.

$9 - 7$	$10 - 6$	$13 - 6$	$8 - 7$	$15 - 7$
$14 - 6$	$7 - 7$	$16 - 7$	$9 - 6$	$11 - 7$
$8 - 6$	$13 - 7$	$12 - 6$	$10 - 7$	$7 - 6$
$15 - 6$	$12 - 7$	$6 - 6$	$11 - 6$	$14 - 7$
$10 - 6$	$14 - 7$	$11 - 6$	$16 - 7$	$8 - 7$

Name ______________________________

Set 19: Subtracting 7 and 6

1. Read the answers to someone.
2. Write the answers.
3. Ask someone to correct your paper. Corrected by ______________________

$14 - 7$	$8 - 6$	$13 - 6$	$7 - 7$	$11 - 6$
$15 - 6$	$8 - 7$	$10 - 6$	$14 - 6$	$12 - 7$
$10 - 7$	$12 - 6$	$16 - 7$	$6 - 6$	$9 - 7$
$15 - 7$	$9 - 6$	$11 - 7$	$7 - 6$	$13 - 7$
$16 - 7$	$10 - 6$	$8 - 7$	$14 - 7$	$11 - 6$

Name •
Draw a 3" line segment.

Date •
Draw a line segment 2" longer than the line segment for your name.

1. Sixteen children in Room 14 at Savin Rock Elementary School wrote pen pal letters. Ten children sent their letters. How many children have letters left to send?

Number sentence ____________________

Answer ____________________

2. What numbers would you use to estimate the sum of 47 and 21? ______ and ______

3. What would be a good estimate of the height of a desk in your classroom?

5 feet 1 inch   1 foot 2 inches   2 feet 1 inch   12 feet 2 inches

4. What fractional part of each shape is shaded?

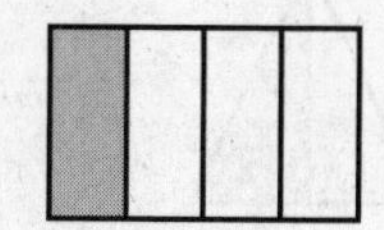 ______

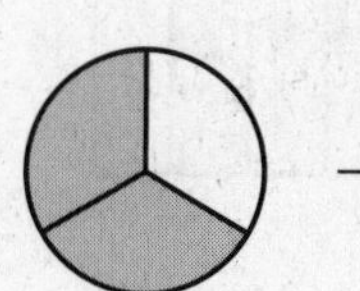 ______

5. How much money is this?
Write the amount two ways.

______ ______

6. Fill in the correct symbol (+ or −).

$3 + 5 = 10 \bigcirc 2$ $10 - 3 = 4 \bigcirc 3$

7. Find the answers.

$\begin{array}{r} 36 \\ -\ 15 \\ \hline \end{array}$ $\begin{array}{r} 50 \\ -\ 27 \\ \hline \end{array}$ $\begin{array}{r} 88 \\ +\ 34 \\ \hline \end{array}$ $\begin{array}{r} 15 \\ 23 \\ +\ 47 \\ \hline \end{array}$

Name ______________________

Date ______________________

Homework 99B

Saxon Math 2 (for use with Lesson 99)

1. Mrs. Roy bought eighteen oranges on March 5th. During the next four days the children ate ten oranges. How many oranges are left?

Number sentence ______________________

Answer ______________________

2. What numbers would you use to estimate the sum of 38 and 51? ______ and ______

3. What would be a good estimate of the height of a car?

5 feet 1 inch   1 foot 2 inches   2 feet 1 inch   12 feet 2 inches

4. What fractional part of each shape is shaded?

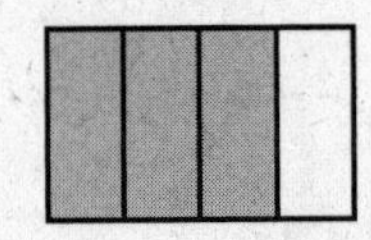 ______

 ______

5. How much money is this?
Write the amount two ways.

______ ______

6. Fill in the correct symbol (+ or –).

2 + 7 = 11 ◯ 2   9 – 1 = 6 ◯ 2

7. Find the answers.

$$\begin{array}{r} 72 \\ -\ 51 \\ \hline \end{array} \qquad \begin{array}{r} 90 \\ -\ 37 \\ \hline \end{array} \qquad \begin{array}{r} 29 \\ +\ 38 \\ \hline \end{array} \qquad \begin{array}{r} 25 \\ 32 \\ +\ 8 \\ \hline \end{array}$$

Name ______________________ Score ________

Fact Assessment 19-1

Saxon *Math 2 (for use with* ***Lesson 100-1****)*

Set 19: Subtracting 7 and 6

$12 - 7$	$9 - 6$	$8 - 7$	$6 - 6$	$11 - 7$
$13 - 6$	$10 - 7$	$14 - 7$	$10 - 6$	$9 - 7$
$12 - 6$	$15 - 7$	$7 - 7$	$11 - 6$	$13 - 7$
$7 - 6$	$9 - 7$	$15 - 6$	$8 - 7$	$10 - 7$
$16 - 7$	$8 - 6$	$11 - 7$	$14 - 6$	$7 - 7$

Name ______________________

Fact Homework 100A

Saxon *Math 2 (for use with* ***Lesson 100-1****)*

Set 19: Subtracting 7 and 6

Pretend you are the teacher.
Correct this paper.
If the answer is incorrect, write the correct answer next to the problem.

12 − 7 5	9 − 6 3	8 − 7 1	6 − 6 0	11 − 7 4
13 − 6 8	10 − 7 3	14 − 7 7	10 − 6 4	9 − 7 2
12 − 6 6	15 − 7 8	7 − 7 0	11 − 6 5	13 − 7 8
7 − 6 1	9 − 7 3	15 − 6 9	8 − 7 1	10 − 7 3
16 − 7 9	8 − 6 2	11 − 7 5	14 − 6 8	7 − 7 0

Name ____________________

Fact Homework 100B

Saxon Math 2 (for use with Lesson 100-1)

S80: 80 Subtraction Facts   Corrected by ________

1	7 − 1	10 − 4	9 − 0	5 − 4	12 − 6	9 − 7	11 − 3	8 − 2	6 − 6	9 − 4
2	6 − 3	11 − 6	10 − 2	6 − 1	2 − 0	9 − 3	7 − 2	6 − 5	5 − 5	11 − 4
3	4 − 2	8 − 0	10 − 6	14 − 5	4 − 0	12 − 7	9 − 5	8 − 4	9 − 2	11 − 5
4	15 − 6	5 − 1	8 − 5	8 − 6	11 − 7	1 − 0	7 − 3	9 − 6	4 − 3	10 − 5
5	12 − 4	13 − 7	8 − 3	16 − 7	10 − 3	4 − 1	6 − 2	13 − 5	7 − 0	11 − 2
6	10 − 7	14 − 6	1 − 1	12 − 3	7 − 5	2 − 1	7 − 7	2 − 2	12 − 5	3 − 1
7	15 − 7	10 − 1	6 − 0	13 − 4	5 − 2	3 − 0	7 − 6	13 − 6	3 − 3	9 − 1
8	6 − 4	7 − 4	8 − 7	4 − 4	3 − 2	5 − 0	5 − 3	14 − 7	0 − 0	8 − 1

Name ______________________________

Lesson Worksheet 100-1A

***Saxon** Math 2 (for use with **Lesson 100-1**)*

A. Write the answers.

8 – 8 = _____

9 – 8 = _____

10 – 8 = _____

11 – 8 = _____

12 – 8 = _____

13 – 8 = _____

14 – 8 = _____

15 – 8 = _____

16 – 8 = _____

17 – 8 = _____

18 – 8 = _____

19 – 8 = _____

B. Draw lines to connect the problems to the answers.

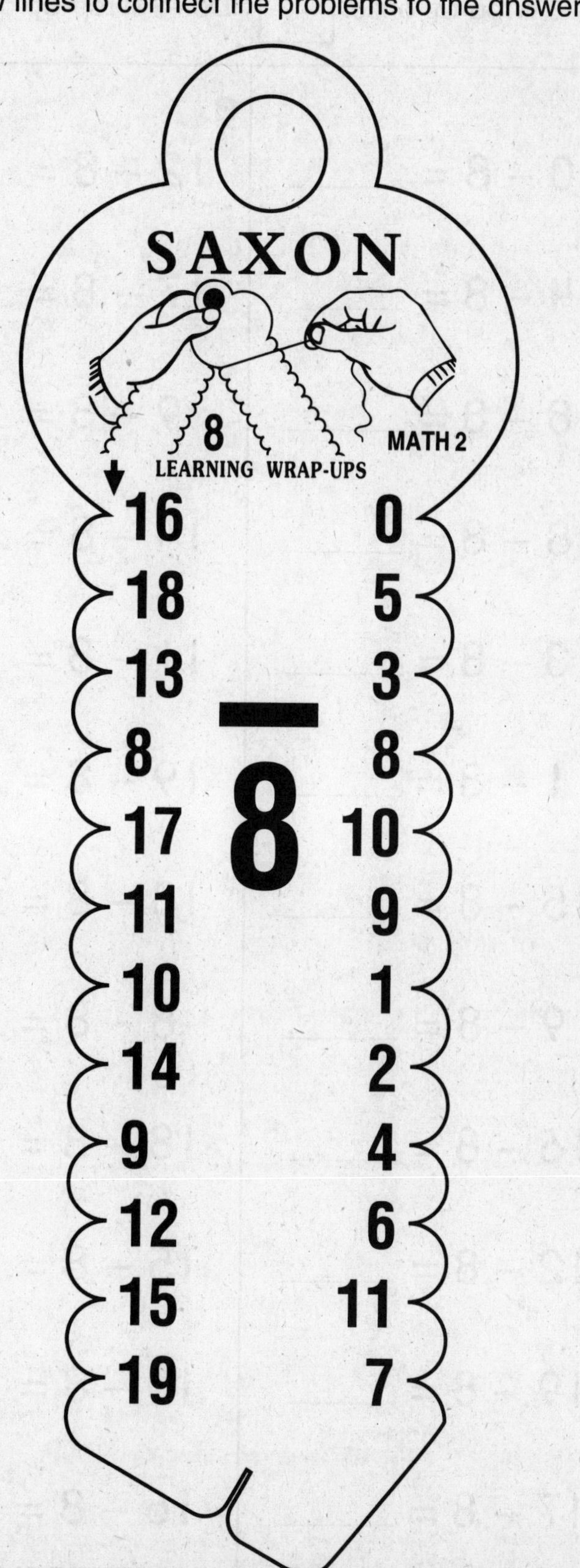

Name ____________________

Saxon Math 2 (for use with Lesson 100-1)

Do –8 Wrap-Up once. ☐ Do –8 Wrap-Up once. ☐ Do –8 Wrap-Up once. ☐

A.

10 – 8 = ____

14 – 8 = ____

8 – 8 = ____

18 – 8 = ____

13 – 8 = ____

11 – 8 = ____

15 – 8 = ____

9 – 8 = ____

16 – 8 = ____

12 – 8 = ____

19 – 8 = ____

17 – 8 = ____

B.

12 – 8 = ____

17 – 8 = ____

9 – 8 = ____

14 – 8 = ____

11 – 8 = ____

19 – 8 = ____

13 – 8 = ____

8 – 8 = ____

18 – 8 = ____

15 – 8 = ____

10 – 8 = ____

16 – 8 = ____

C.

13 – 8 = ____

18 – 8 = ____

10 – 8 = ____

17 – 8 = ____

14 – 8 = ____

9 – 8 = ____

11 – 8 = ____

16 – 8 = ____

12 – 8 = ____

19 – 8 = ____

15 – 8 = ____

8 – 8 = ____

Name ______________________________

Date ______________________________

Problem-Solving Worksheet 100A

***Saxon** Math 2 (for use with **Lesson 100-1**)*

Understand	Plan	Solve	Check

Make a Table 

Look for a Pattern 

The children in Ms. Kaim's class are making books. The children will use 5 pieces of writing paper for each book. Show how many pieces of writing paper Victoria will need to make 6 books.

Number of books	1	2				
Number of pieces of writing paper	5					

How many pieces of writing paper will Victoria need to make 6 books? __________

Name ______________________

Performance Task Worksheet 100B

***Saxon** Math 2 (for use with **Lesson 100-1**)*

Understand	Plan	Solve	Check

The children in Mrs. Klassen's class are making greeting cards with moveable eyes. The children will use 2 moveable eyes for each card. Show how many moveable eyes Tyrise will need to make 6 greeting cards.

Number of greeting cards	1	2				
Number of moveable eyes	2					

How many moveable eyes will Tyrise need to make 6 greeting cards? ____________

Circle the problem-solving strategies you used to solve this problem.

Act It Out

Draw a Picture

Make an Organized List

Make a Table

Use Logical Reasoning

Look for a Pattern

Guess and Check

Explain how you got your answer: ______________________________

____

____

____

Name ______________________ Score ________

Fact Assessment 19-2

***Saxon** Math 2 (for use with **Lesson 100-2**)*

S80: 80 Subtraction Facts

7 − 1	10 − 4	9 − 0	5 − 4	12 − 6	9 − 7	11 − 3	8 − 2	6 − 6	9 − 4
6 − 3	11 − 6	10 − 2	6 − 1	2 − 0	9 − 3	7 − 2	6 − 5	5 − 5	11 − 4
4 − 2	8 − 0	10 − 6	14 − 5	4 − 0	12 − 7	9 − 5	8 − 4	9 − 2	11 − 5
15 − 6	5 − 1	8 − 5	8 − 6	11 − 7	1 − 0	7 − 3	9 − 6	4 − 3	10 − 5
12 − 4	13 − 7	8 − 3	16 − 7	10 − 3	4 − 1	6 − 2	13 − 5	7 − 0	11 − 2
10 − 7	14 − 6	1 − 1	12 − 3	7 − 5	2 − 1	7 − 7	2 − 2	12 − 5	3 − 1
15 − 7	10 − 1	6 − 0	13 − 4	5 − 2	3 − 0	7 − 6	13 − 6	3 − 3	9 − 1
6 − 4	7 − 4	8 − 7	4 − 4	3 − 2	5 − 0	5 − 3	14 − 7	0 − 0	8 − 1

Name ______________________

Lesson Worksheet 100-2

***Saxon** Math 2 (for use with **Lesson 100-2**)*

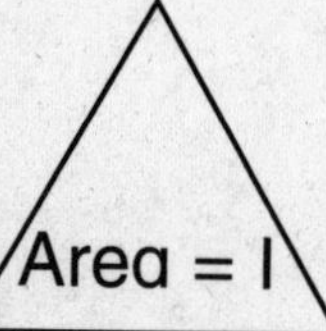

Area = ______

Area = ______

Area = ______

A.

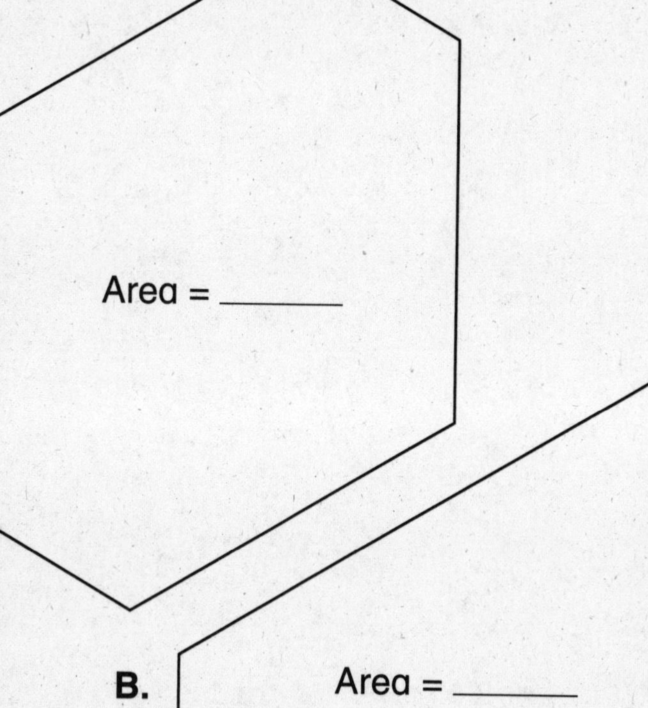

B. Area = ______

C.

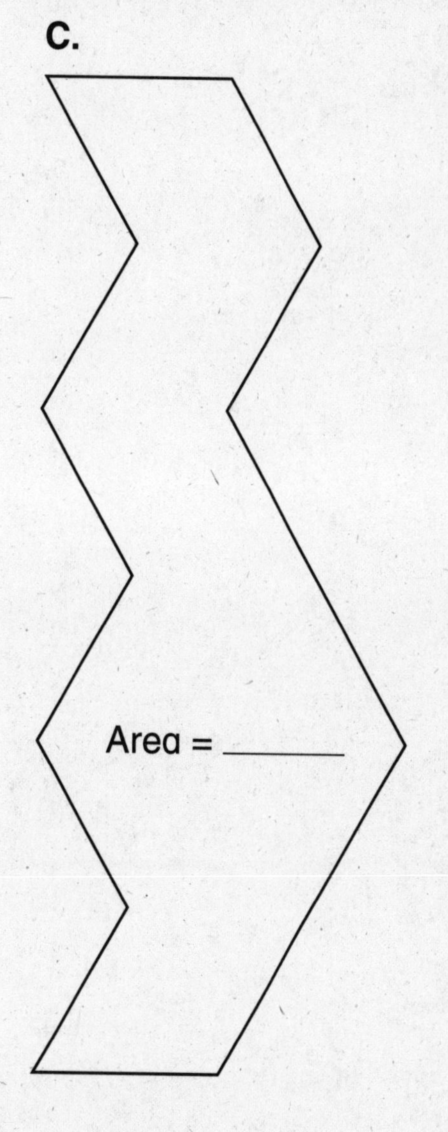

Name ______________________________

Date ______________________________

1. James has 3 dimes and 7 pennies. George has 8 pennies and 4 dimes. How much money does each boy have?

James __________ George __________

How much money do they have altogether?

Number sentence ______________________________

Answer ______________________________

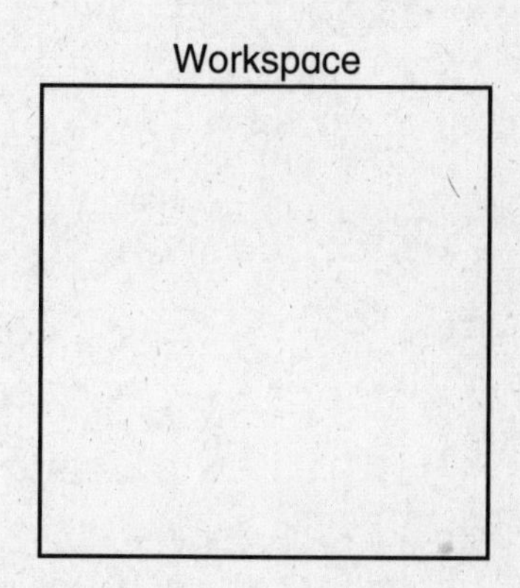

2. One of these names is the name of my dog. Use the clues to find the name of my dog. Cross out the names that cannot be my dog's name.

It is not the name that is last.
It does not have exactly 4 letters.
It is not the name in the middle.

What is my dog's name?

____

Lady	Duffer	Rover	Spot	Ebony

3. Show 4:35 on the clock.

4. Color five circles red.

What fractional part of the circles is red? ____

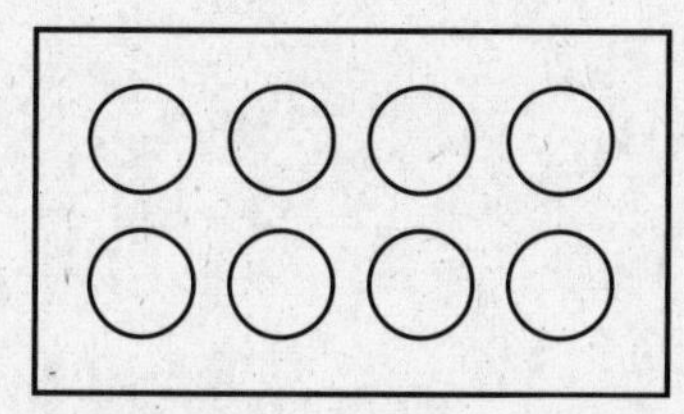

5. How many children were absent on Monday? ______

How many children were absent on Tuesday? ______

Draw pictures to show that 6 children were absent on Wednesday.

Absent Children

Mon.	☺ ☺ ☺ ☺
Tues.	☺ ◖
Wed.	

☺ = 2 children

6. Find the differences.

$$\begin{array}{r} 54 \\ -\ 15 \\ \hline \end{array} \qquad \begin{array}{r} 72 \\ -\ 34 \\ \hline \end{array} \qquad \begin{array}{r} 68 \\ -\ 23 \\ \hline \end{array} \qquad \begin{array}{r} 80 \\ -\ 31 \\ \hline \end{array}$$

Name ________________________________

Class Fact Practice 101A

***Saxon** Math 2 (for use with **Lesson 101**)*

Set 20: Subtracting 8 and 7; Review Facts

Do –8 Wrap-Up once.
Do –7 Wrap-Up once.

14 − 8	8 − 8	12 − 8	16 − 8	10 − 8
17 − 8	11 − 8	15 − 8	9 − 8	13 − 8
10 − 7	16 − 7	7 − 7	15 − 7	11 − 7
13 − 7	9 − 7	14 − 7	8 − 7	12 − 7
10 − 6	9 − 6	6 − 6	14 − 6	12 − 6

Name ____________________

Saxon Math 2 (for use with Lesson 101)

Set 20: Subtracting 8 and 7; Review Facts

1. Read the answers to someone.
2. Write the answers.
3. Ask someone to correct your paper. Corrected by ____________________

13 − 8	8 − 8	16 − 8	12 − 8	10 − 8
15 − 8	11 − 8	14 − 8	9 − 8	17 − 8
11 − 7	15 − 7	7 − 7	13 − 7	10 − 7
12 − 7	8 − 7	16 − 7	14 − 7	9 − 7
11 − 6	10 − 6	9 − 6	12 − 6	7 − 6

Name ______________________________

***Saxon** Math 2 (for use with **Lesson 101**)*

Cone	Cube	Sphere
Cylinder	**Rectangular Prism**	**Pyramid**

Name •————————•

Measure this line segment using inches. ______"

Guided Class Practice 101A

***Saxon** Math 2 (for use with **Lesson 101**)*

Date •

Draw a line segment 1" longer than the line segment for your name.

1. Lia had 50 paper clips. She gave 18 paper clips to Anna. How many paper clips does Lia have now?

Number sentence ______________________

Answer ______________________

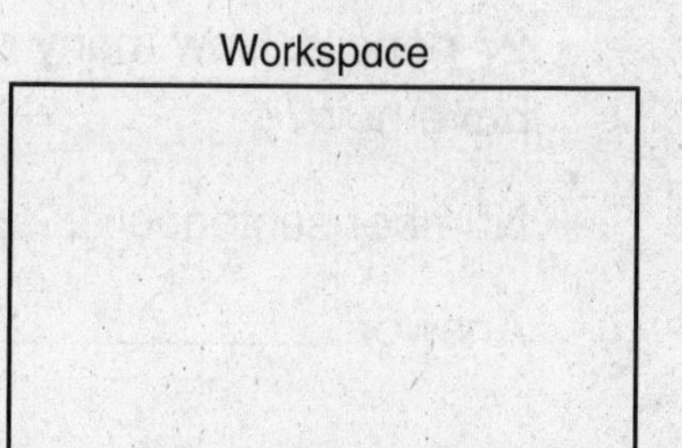

2. Circle the best number sentence to use to estimate the sum of 51 and 38.

50 + 30 = 80   50 + 40 = 90   60 + 30 = 90   60 + 40 = 100

3. Show how two children will share the markers equally.

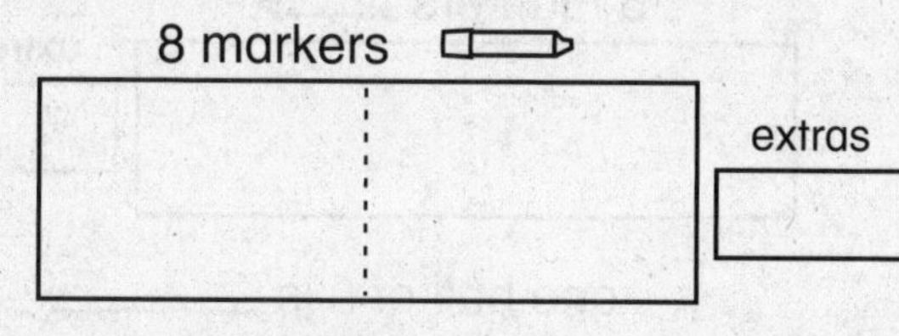

one half of 8 is ______

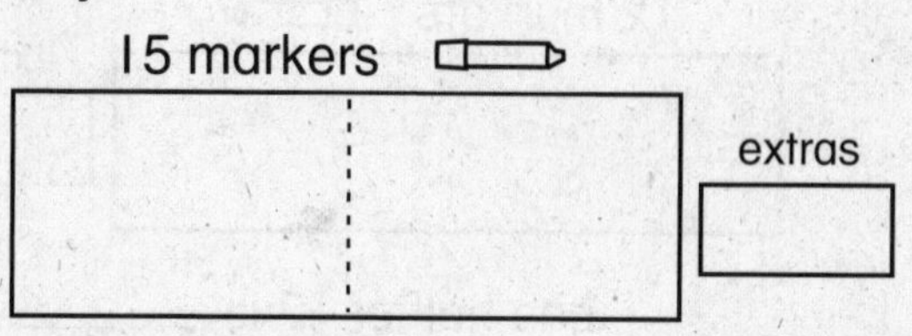

one half of 15 is ______

4. Circle the name of this geometric solid.

cone   rectangular prism   pyramid   cylinder

5. On what day were the fewest children absent? ______________

On what day were fewer children absent than the day before? ______________

How many days were the same number of children absent? ________

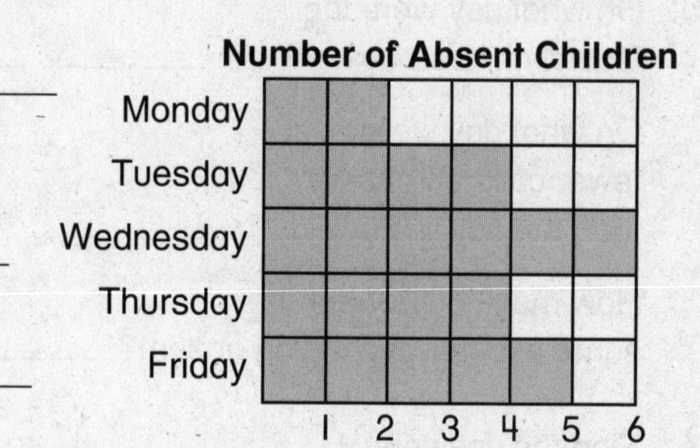

6. Find the answers.

$$\begin{array}{r} 65 \\ -\ 18 \\ \hline \end{array} \qquad \begin{array}{r} 90 \\ -\ 38 \\ \hline \end{array} \qquad \begin{array}{r} 38 \\ +\ 91 \\ \hline \end{array}$$

$4 \times 10 =$ ______ $6 \times 10 =$ ______

Name ______________________

Date ______________________

Homework 101B

***Saxon** Math 2 (for use with **Lesson 101**)*

1. Larry has 40 colored pencils. He gave Barbara 27 pencils. How many colored pencils does Larry have now?

Number sentence ______________________

Answer ______________________

Workspace

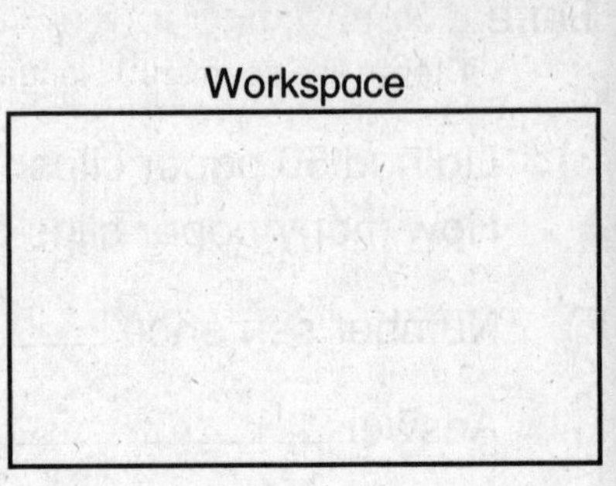

2. Circle the best number sentence to use to estimate the sum of 29 and 48.

$20 + 40 = 60$ $20 + 50 = 70$ $30 + 40 = 70$ $30 + 50 = 80$

3. Show how two children will share the markers equally.

12 markers

extras

one half of 12 is ______

5 markers

extras

one half of 5 is ______

4. Circle the name of this geometric solid.

cone rectangular prism pyramid cylinder

5. On what day were the fewest children absent? ______________

On what day were fewer children absent than the day before? ______________

How many days were the same number of children absent? ______

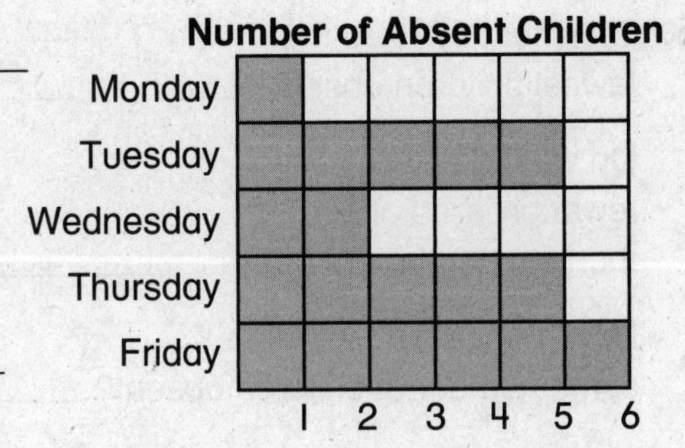

6. Find the answers.

$$\begin{array}{r} 72 \\ -\ 28 \\ \hline \end{array} \qquad \begin{array}{r} 89 \\ -\ 58 \\ \hline \end{array} \qquad \begin{array}{r} 72 \\ +\ 83 \\ \hline \end{array}$$

$8 \times 10 =$ ______ $3 \times 10 =$ ______

Name ______________________________

Class Fact Practice 102A

***Saxon** Math 2 (for use with **Lesson 102**)*

Set 20: Subtracting 8 and 7

Do –8 Wrap-Up twice.	Do –7 Wrap-Up twice.
13 – 8 = ____	10 – 7 = ____
11 – 8 = ____	16 – 7 = ____
8 – 8 = ____	9 – 7 = ____
15 – 8 = ____	13 – 7 = ____
12 – 8 = ____	8 – 7 = ____
9 – 8 = ____	14 – 7 = ____
14 – 8 = ____	7 – 7 = ____
17 – 8 = ____	12 – 7 = ____
10 – 8 = ____	15 – 7 = ____
16 – 8 = ____	11 – 7 = ____

Name ________________________________

Saxon Math 2 (for use with **Lesson 102**)

Set 20: Subtracting 8 and 7

1. Write the answers.
2. Draw lines to connect the problems that have the same answer.
3. Ask someone to correct your paper. Corrected by ______________________

10 – 8 = _____ •	• 16 – 7 = _____
14 – 8 = _____ •	• 8 – 7 = _____
17 – 8 = _____ •	• 15 – 7 = _____
9 – 8 = _____ •	• 9 – 7 = _____
16 – 8 = _____ •	• 7 – 7 = _____
12 – 8 = _____ •	• 13 – 7 = _____
8 – 8 = _____ •	• 11 – 7 = _____
15 – 8 = _____ •	• 10 – 7 = _____
11 – 8 = _____ •	• 12 – 7 = _____
13 – 8 = _____ •	• 14 – 7 = _____

Name

Draw an 8-cm line segment.

Date

Measure this line segment using centimeters. ______ cm

Guided Class Practice 102A

*Saxon Math 2 (for use with **Lesson 102**)*

1. There are 25 children in Room 12. Eighteen of those children chose math as their favorite subject. There are 24 children in Room 14. Nineteen of those children chose math as their favorite subject. Altogether, how many children chose math as their favorite subject?

Number sentence ______________________

Answer ______________________

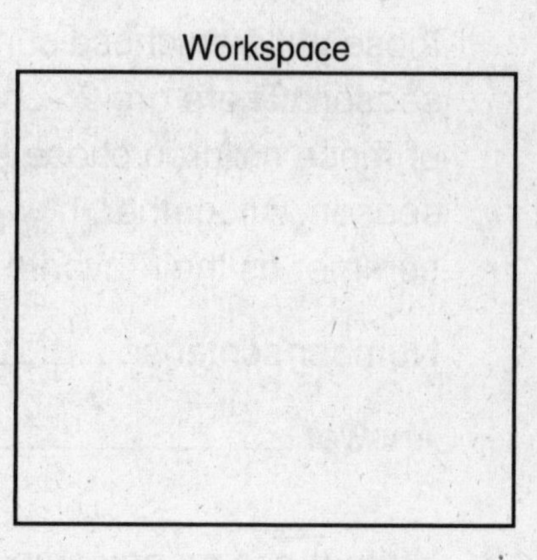

2. Write these money amounts in order from least to greatest. Circle the money amount that is between one dollar and two dollars.

$1.79	$3.89	87¢	$2.03

______ least ______ ______ ______ greatest

3. Round each number to the nearest 10.

19 ______ 95 ______ 52 ______

4. Fill in the missing numbers in these number patterns.

______, ______, ______, 100, 110, 120, ______, ______, ______

______, ______, ______, 18, 17, 16, ______, ______, ______

5. Fill in the missing symbol (+, –, or ×).

$5 + 2 = 4 \bigcirc 3$ ______ $2 \bigcirc 10 = 20$ $9 - 4 = 8 \bigcirc 3$ ______

6. Find the answers.

81 – 38

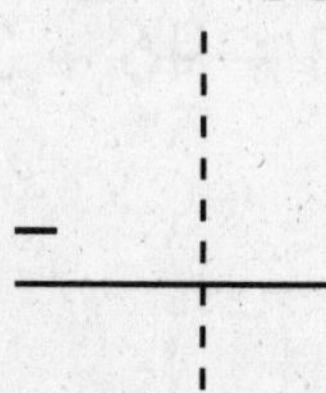

79 – 25

68 + 7 + 13

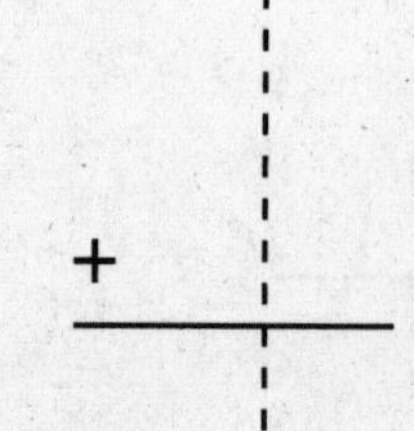

29 + 73 + 21

Name ______________________

Date ______________________

Saxon Math 2 (for use with Lesson 102)

1. There are 28 children in Room 17. Twenty-one of those children chose summer as their favorite season. There are 26 children in Room 18. Nineteen of those children chose summer as their favorite season. Altogether, how many children chose summer as their favorite season?

Number sentence ______________________

Answer ______________________

Workspace

2. Write these money amounts in order from least to greatest. Circle the money amount that is between one dollar and two dollars.

$2.10	95¢	$2.75	$1.62

______ ______ ______ ______
least   greatest

3. Round each number to the nearest 10.

47 ______   93 ______   15 ______

4. Fill in the missing numbers in these number patterns.

______, ______, ______, 200, 210, 220, ______, ______, ______

______, ______, ______, 26, 25, 24, ______, ______, ______

5. Fill in the missing symbol (+, –, or ×).

$7 + 10 = 9 \bigcirc 8$ $4 + 3 = 10$  3 $3 \bigcirc 10 = 30$

6. Find the answers.

84 – 65   89 – 48   24 + 8 + 56   73 + 46 + 31

–

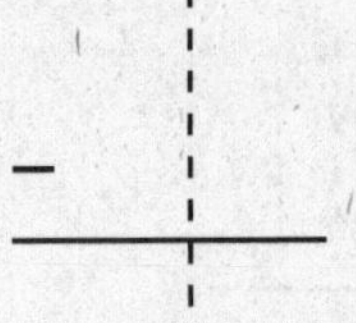

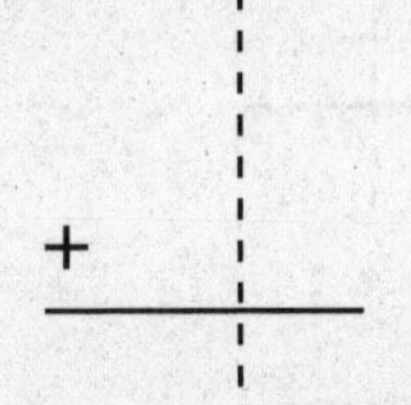

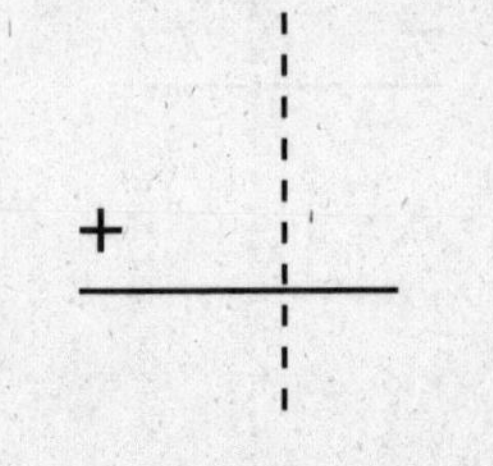

Name ____________________

Class Fact Practice 103A

***Saxon** Math 2 (for use with **Lesson 103**)*

Set 20: Subtracting 8 and 7

8 − 8	13 − 8	14 − 7	17 − 8	10 − 7
15 − 8	12 − 8	8 − 7	14 − 8	11 − 7
9 − 8	16 − 8	13 − 7	11 − 8	15 − 7
17 − 8	12 − 7	16 − 7	13 − 8	16 − 8
9 − 7	14 − 8	10 − 8	7 − 7	15 − 8

Name ______________________________

Saxon Math 2 (for use with **Lesson 103**)

Set 20: Subtracting 8 and 7

1. Read the answers to someone.
2. Write the answers.
3. Ask someone to correct your paper. Corrected by ______________________

15 − 8	7 − 7	10 − 8	14 − 8	9 − 7
16 − 8	13 − 8	16 − 7	12 − 7	17 − 8
15 − 7	11 − 8	13 − 7	16 − 8	9 − 8
11 − 7	14 − 8	8 − 7	12 − 8	15 − 8
10 − 7	17 − 8	14 − 7	13 − 8	8 − 8

Name

Draw a 7-cm line segment.

Guided Class Practice 103A

***Saxon** Math 2 (for use with **Lesson 103**)*

Date

Measure this line segment using centimeters. ______ cm

1. Mary bought 6 packages of candy. There are 10 candies in a package. How many candies does she have?

Number sentence ______

Answer ______

2. Draw a picture to show one hundred thirteen. (Use ▭ for 100, ▯ for 10, and ▫ for 1.)

Write the number in expanded form. ______

Circle the number that shows one hundred thirteen.

10013 113 1013 100103

3. Color the cone yellow.
Color the pyramid blue.
Color the rectangular prism red.

  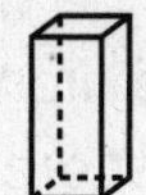 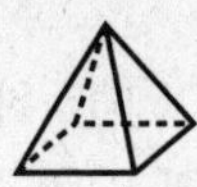 

4. I have 10 quarters. Draw the coins. How much money do I have? ______

5. About how long is a new pencil?

1 foot 2 inches 1 inch 18 inches 7 inches

6. Find the answers.

$3 \times 100 =$ ______

$2 \times 1 =$ ______

$6 \times 10 =$ ______

$$\begin{array}{r} 28 \\ 37 \\ +\ 53 \\ \hline \end{array} \qquad \begin{array}{r} 92 \\ -\ 38 \\ \hline \end{array} \qquad \begin{array}{r} 63 \\ -\ 57 \\ \hline \end{array}$$

Name ______________________

Date ______________________

1. Silvia bought a package of pencils. There are 10 pencils in a package. Silvia gave 4 pencils to her sister. How many pencils does she have left?

Number sentence ______________________

Answer ______________________

2. Draw a picture to show three hundred twenty-four.

(Use ☐ for 100, ▯ for 10, and ▫ for 1.)

Write the number in expanded form. ______________________

Circle the number that shows three hundred twenty-four.

3204   324   30024   300204

3. Color the sphere orange.
Color the cylinder green.
Color the cube purple.

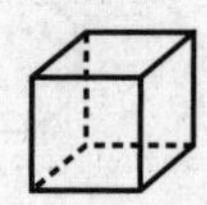

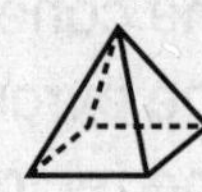

4. I have 8 quarters. Draw the coins. How much money do I have? __________

5. About how long is a new crayon?

2 feet   12 inches   1 inch   4 inches

6. Find the answers.

$5 \times 1 =$ ______

$9 \times 100 =$ ______

$7 \times 10 =$ ______

$$\begin{array}{r} 32 \\ 58 \\ +\ 63 \\ \hline \end{array} \qquad \begin{array}{r} 45 \\ -\ 17 \\ \hline \end{array} \qquad \begin{array}{r} 80 \\ -\ 38 \\ \hline \end{array}$$

Name ______________________________

Class Fact Practice 104A

***Saxon** Math 2 (for use with **Lesson 104**)*

Set 20: Subtracting 8 and 7

Do –8 Wrap-Up twice.
Do –7 Wrap-Up once.

12 − 8	7 − 7	15 − 7	11 − 8	16 − 8
14 − 7	10 − 8	12 − 7	8 − 7	13 − 8
10 − 7	15 − 8	16 − 7	9 − 8	11 − 7
13 − 7	8 − 8	17 − 8	9 − 7	14 − 8
14 − 8	9 − 7	10 − 8	16 − 8	12 − 7

Name ______________________________

Saxon Math 2 (for use with Lesson 104)

Set 20: Subtracting 8 and 7

1. Read the answers to someone.
2. Write the answers.
3. Ask someone to correct your paper. Corrected by ____________________

13 − 7	9 − 8	13 − 8	7 − 7	16 − 7
17 − 8	10 − 7	12 − 8	15 − 8	11 − 7
15 − 7	16 − 8	8 − 7	10 − 8	14 − 7
11 − 8	12 − 7	14 − 8	8 − 8	9 − 7
9 − 7	14 − 8	16 − 8	12 − 7	10 − 8

Name ______________________

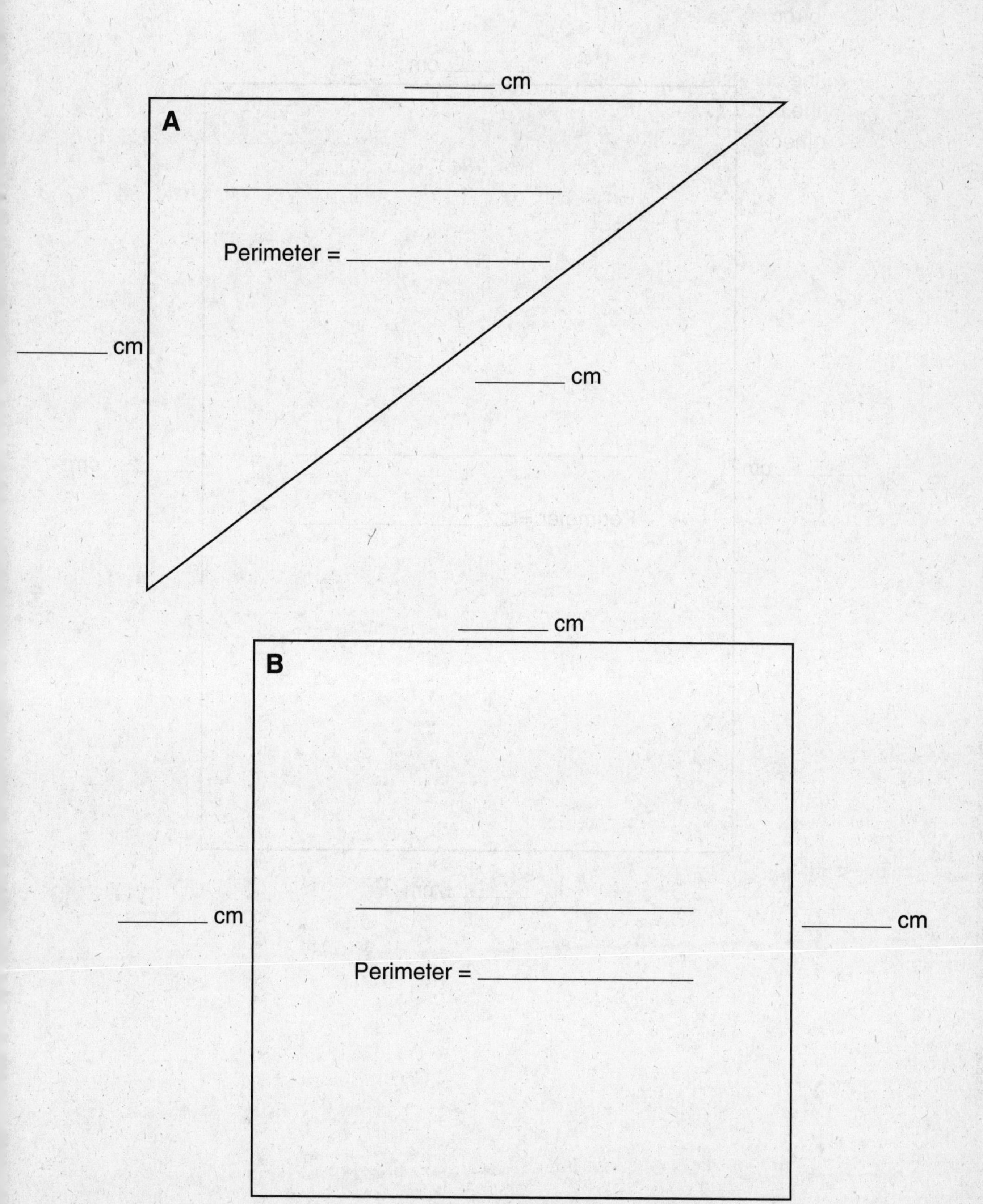

Name ______________________________

_______ cm

C

_______ cm

_______ cm

____

Perimeter = ____________________

_______ cm

Name •
Draw an 8-cm line segment.

Date •———————•
Measure this line segment using centimeters. _______ cm

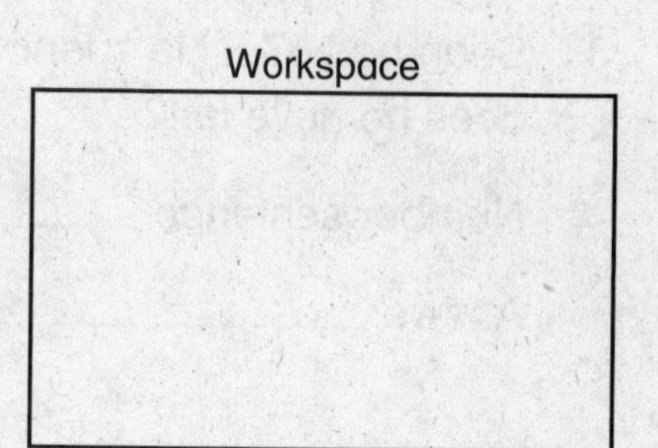

1. Olivia had 78¢. She spent 29¢. How much money does she have left?

Number sentence ____________

Answer ____________

2. Show how to divide the pennies in half.

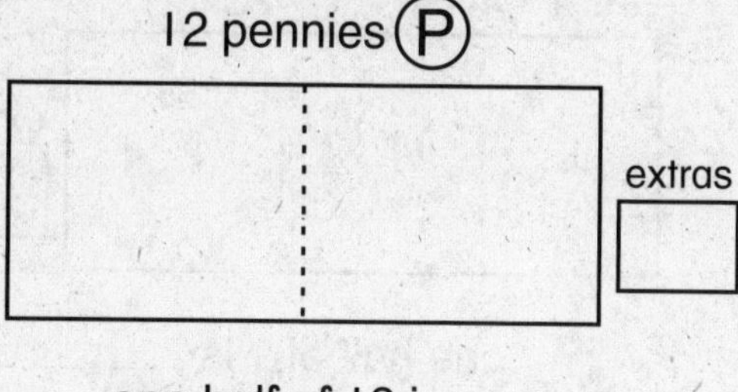

one half of 12 is _____

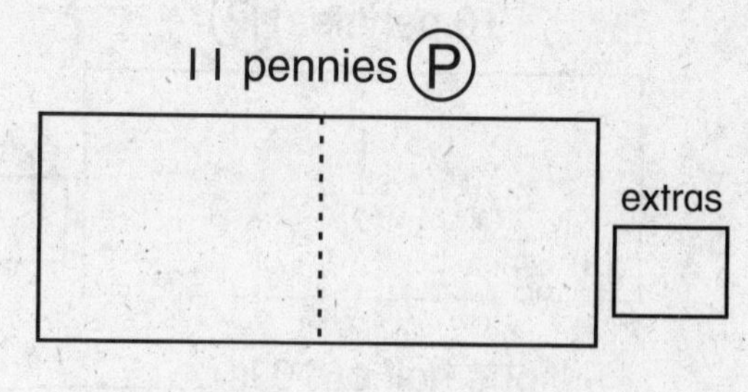

one half of 11 is _____

3. Use the correct comparison symbol (>, <, or =).

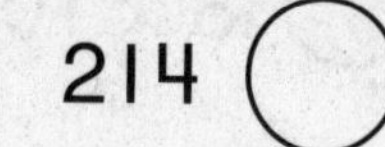

214 ◯ 341 15 – 8 ◯ 14 – 7 28 + 10 ◯ 3 × 10

4. Measure and label each side of the triangle using centimeters. What is the perimeter?

Number sentence ____________

Perimeter ____________

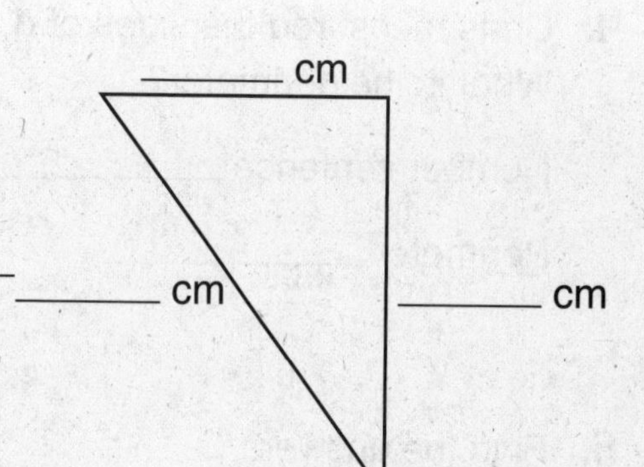

5. Find the answers.

$$\begin{array}{r} 84 \\ -\ 38 \\ \hline \end{array} \qquad \begin{array}{r} 25 \\ -\ 17 \\ \hline \end{array}$$

$8 \times 100 =$ _____

$2 \times 1 =$ _____

$$\begin{array}{r} 51 \\ 19 \\ +\ 23 \\ \hline \end{array} \qquad \begin{array}{r} 3 \\ 7 \\ 2 \\ 4 \\ 8 \\ +\ 5 \\ \hline \end{array}$$

6. Fill in the missing numbers in this number pattern.

292, 294, 296, _____, _____, _____, _____, _____, _____

Name ____________________

Date ____________________

***Saxon** Math 2 (for use with **Lesson 104**)*

1. Sean had 87¢. He spent 39¢. How much money does he have left?

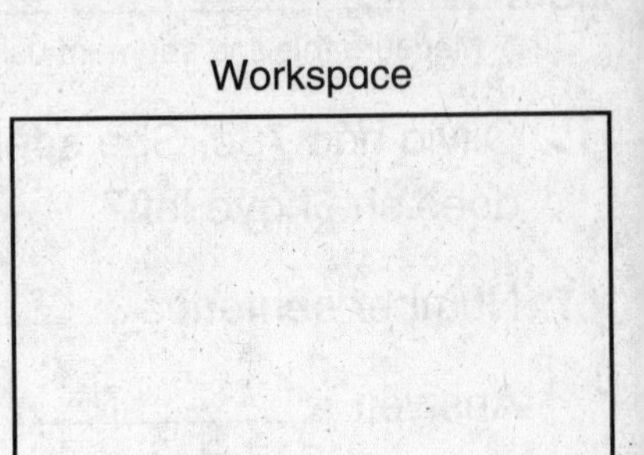

Number sentence ____________________

Answer ____________________

2. Show how to divide the pennies in half.

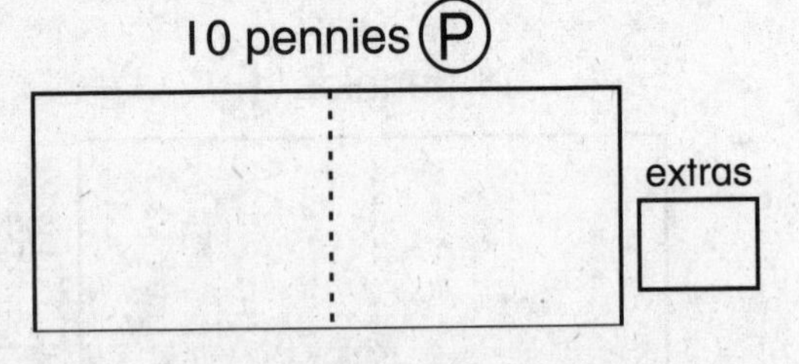

one half of 10 is ______

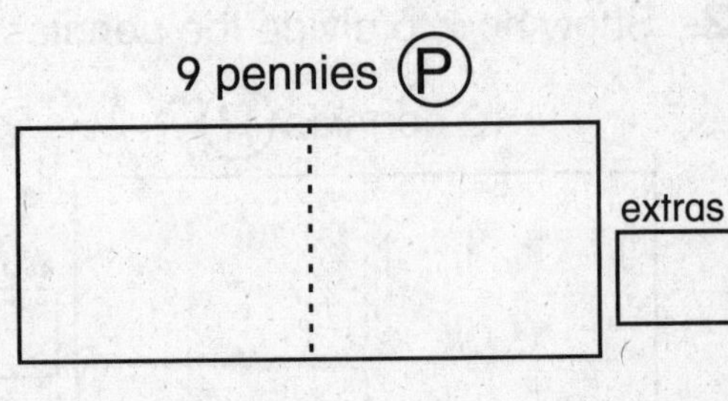

one half of 9 is ______

3. Use the correct comparison symbol (>, <, or =).

638 ◯ 429   17 − 8 ◯ 15 − 7   39 + 10 ◯ 5 × 10

______ ______   ______ ______

4. Chris measured the sides of a triangle and drew a small picture. What is the perimeter?

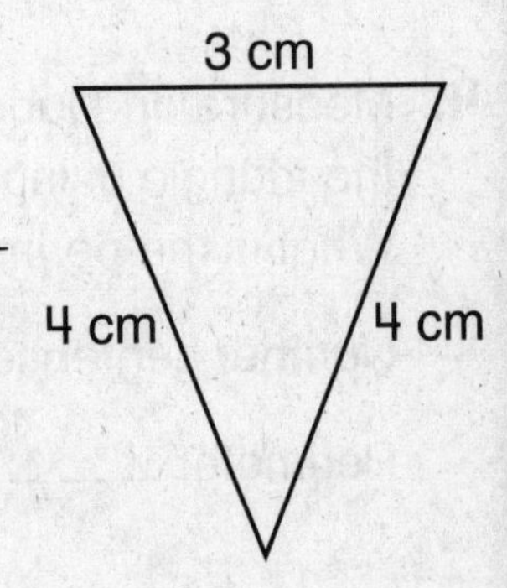

Number sentence ____________________

Perimeter ______

5. Find the answers.

7 × 100 = ______

3 × 1 = ______

$$\begin{array}{r} 46 \\ 23 \\ +\ 16 \\ \hline \end{array}$$

$$\begin{array}{r} 3 \\ 1 \\ 7 \\ 8 \\ 6 \\ +\ 2 \\ \hline \end{array}$$

6. Fill in the missing numbers in this number pattern.

293, 295, 297, ______, ______, ______, ______, ______, ______

Name ______________________ Score ______

Fact Assessment 20-1

***Saxon** Math 2 (for use with **Lesson 105-1**)*

Set 20: Subtracting 8 and 7

$9 - 8$	$15 - 8$	$12 - 8$	$8 - 7$	$10 - 8$
$17 - 8$	$11 - 7$	$8 - 8$	$13 - 8$	$9 - 7$
$11 - 8$	$16 - 7$	$14 - 8$	$9 - 8$	$12 - 7$
$16 - 8$	$14 - 7$	$8 - 8$	$13 - 7$	$10 - 8$
$12 - 8$	$10 - 7$	$11 - 8$	$15 - 7$	$7 - 7$

Name ____________________

Fact Homework **105A**

***Saxon** Math 2 (for use with **Lesson 105-1**)*

Set 20: Subtracting 8 and 7

Pretend you are the teacher.
Correct this paper.
If the answer is incorrect, write the correct answer next to the problem.

$\begin{array}{r} 9 \\ -\ 8 \\ \hline 1 \end{array}$	$\begin{array}{r} 15 \\ -\ 8 \\ \hline 7 \end{array}$	$\begin{array}{r} 12 \\ -\ 8 \\ \hline 4 \end{array}$	$\begin{array}{r} 8 \\ -\ 7 \\ \hline 1 \end{array}$	$\begin{array}{r} 10 \\ -\ 8 \\ \hline 2 \end{array}$
$\begin{array}{r} 17 \\ -\ 8 \\ \hline 9 \end{array}$	$\begin{array}{r} 11 \\ -\ 7 \\ \hline 3 \end{array}$	$\begin{array}{r} 8 \\ -\ 8 \\ \hline 0 \end{array}$	$\begin{array}{r} 13 \\ -\ 8 \\ \hline 5 \end{array}$	$\begin{array}{r} 9 \\ -\ 7 \\ \hline 2 \end{array}$
$\begin{array}{r} 11 \\ -\ 8 \\ \hline 3 \end{array}$	$\begin{array}{r} 16 \\ -\ 7 \\ \hline 8 \end{array}$	$\begin{array}{r} 14 \\ -\ 8 \\ \hline 6 \end{array}$	$\begin{array}{r} 9 \\ -\ 8 \\ \hline 1 \end{array}$	$\begin{array}{r} 12 \\ -\ 7 \\ \hline 6 \end{array}$
$\begin{array}{r} 16 \\ -\ 8 \\ \hline 8 \end{array}$	$\begin{array}{r} 14 \\ -\ 7 \\ \hline 7 \end{array}$	$\begin{array}{r} 8 \\ -\ 8 \\ \hline 0 \end{array}$	$\begin{array}{r} 13 \\ -\ 7 \\ \hline 6 \end{array}$	$\begin{array}{r} 10 \\ -\ 8 \\ \hline 2 \end{array}$
$\begin{array}{r} 12 \\ -\ 8 \\ \hline 5 \end{array}$	$\begin{array}{r} 10 \\ -\ 7 \\ \hline 3 \end{array}$	$\begin{array}{r} 11 \\ -\ 8 \\ \hline 3 \end{array}$	$\begin{array}{r} 15 \\ -\ 7 \\ \hline 8 \end{array}$	$\begin{array}{r} 7 \\ -\ 7 \\ \hline 0 \end{array}$

Name ____________________

Fact Homework 105B

***Saxon** Math 2 (for use with **Lesson 105-1**)*

S90: 90 Subtraction Facts   Corrected by ________

1	7 − 1	10 − 4	9 − 0	5 − 4	12 − 6	9 − 7	11 − 3	8 − 2	6 − 6	9 − 4
2	6 − 3	11 − 6	10 − 2	6 − 1	12 − 8	2 − 0	9 − 3	7 − 2	6 − 5	5 − 5
3	11 − 4	4 − 2	8 − 0	10 − 6	14 − 5	4 − 0	12 − 7	9 − 5	8 − 4	13 − 8
4	9 − 2	11 − 5	15 − 6	5 − 1	8 − 5	16 − 8	8 − 6	11 − 7	1 − 0	7 − 3
5	9 − 6	4 − 3	17 − 8	10 − 5	12 − 4	13 − 7	8 − 3	16 − 7	10 − 3	4 − 1
6	6 − 2	13 − 5	7 − 0	11 − 2	10 − 8	10 − 7	14 − 6	1 − 1	12 − 3	7 − 5
7	2 − 1	11 − 8	7 − 7	2 − 2	12 − 5	3 − 1	15 − 7	10 − 1	6 − 0	13 − 4
8	5 − 2	9 − 8	3 − 0	7 − 6	13 − 6	3 − 3	14 − 8	9 − 1	6 − 4	7 − 4
9	8 − 7	4 − 4	15 − 8	3 − 2	5 − 0	5 − 3	8 − 8	14 − 7	0 − 0	8 − 1

Name ______________________

Lesson Worksheet 105-1A

***Saxon** Math 2 (for use with **Lesson 105-1**)*

A. Write the answers.

$9 - 9 =$ ______

$10 - 9 =$ ______

$11 - 9 =$ ______

$12 - 9 =$ ______

$13 - 9 =$ ______

$14 - 9 =$ ______

$15 - 9 =$ ______

$16 - 9 =$ ______

$17 - 9 =$ ______

$18 - 9 =$ ______

$19 - 9 =$ ______

$20 - 9 =$ ______

B. Draw lines to connect the problems to the answers.

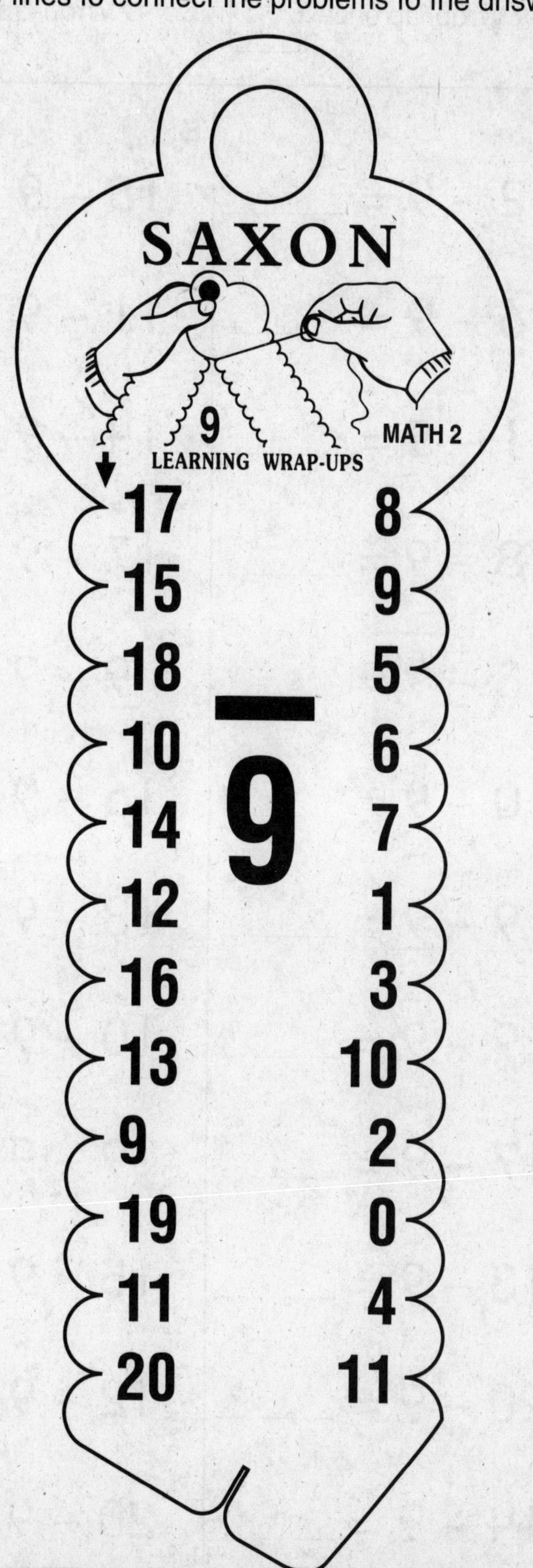

Name ______________________________

Saxon Math 2 (for use with **Lesson 105-1**)

Do −9 Wrap-Up once. ☐ Do −9 Wrap-Up once. ☐ Do −9 Wrap-Up once. ☐

A.	B.	C.
12 − 9 = ____	18 − 9 = ____	11 − 9 = ____
16 − 9 = ____	11 − 9 = ____	17 − 9 = ____
11 − 9 = ____	14 − 9 = ____	9 − 9 = ____
18 − 9 = ____	17 − 9 = ____	13 − 9 = ____
15 − 9 = ____	9 − 9 = ____	16 − 9 = ____
10 − 9 = ____	16 − 9 = ____	12 − 9 = ____
19 − 9 = ____	13 − 9 = ____	19 − 9 = ____
9 − 9 = ____	10 − 9 = ____	14 − 9 = ____
17 − 9 = ____	19 − 9 = ____	18 − 9 = ____
13 − 9 = ____	15 − 9 = ____	10 − 9 = ____
20 − 9 = ____	12 − 9 = ____	20 − 9 = ____
14 − 9 = ____	20 − 9 = ____	15 − 9 = ____

Name
Draw a 7-cm line segment.

Date

Measure this line segment using centimeters. ________ cm

Guided Class Practice 105A

***Saxon** Math 2 (for use with **Lesson 105-1**)*

1. Leah said that whenever she adds two odd numbers the answer is always an even number. Add 3 pairs of odd numbers to see if she is right.

☐ + ☐ = ______ ☐ + ☐ = ______ ☐ + ☐ = ______
odd odd ______ odd odd ______ odd odd ______

2. Show how two children will share the markers equally.

13 markers — extras

one half of 13 is ______

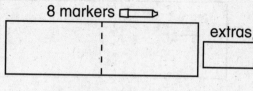

8 markers — extras

one half of 8 is ______

3. Show eight twenty-five on the clocks.

☐ :

______ cm

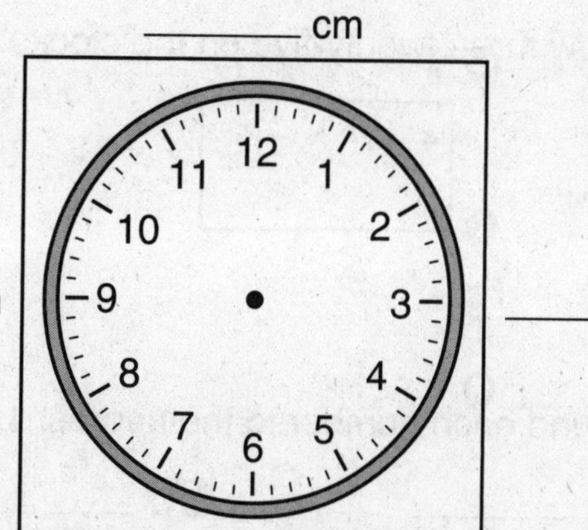

______ cm

______ cm

______ cm

4. Round each number to the nearest 10.

73 ______ 59 ______

18 ______ 35 ______

5. Measure and label each side of the square around the clock.
What is the perimeter of the square?

Number sentence ______________________

Perimeter ______________

6. Find the answers.

53	53	62	62
+ 27	− 27	+ 39	− 39

Name ____________________

Date ____________________

Saxon Math 2 (for use with **Lesson 105-1**)

1. Martell said that whenever he adds two even numbers the answer is always an even number. Add 3 pairs of even numbers to see if he is right.

☐ + ☐ = ______ ☐ + ☐ = ______ ☐ + ☐ = ______

even even ______ even even ______ even even ______

2. Show how two children will share the pencils equally.

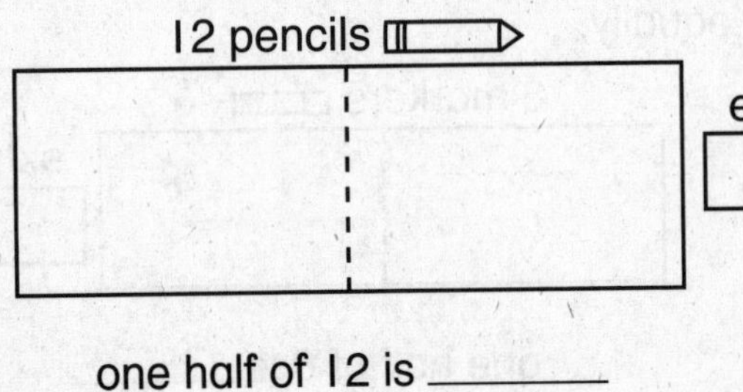

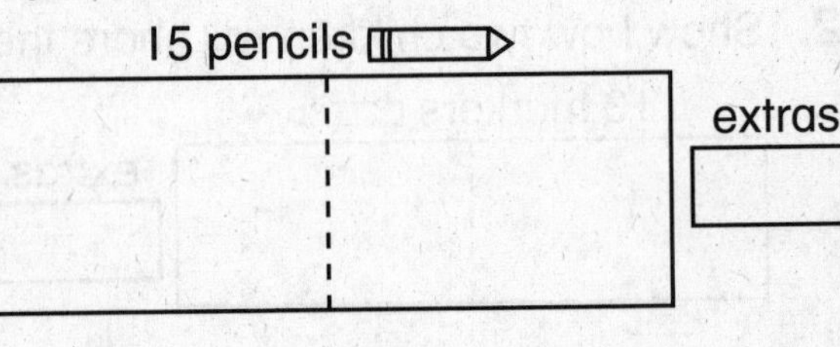

one half of 12 is ______ one half of 15 is ______

3. Show three twenty-five on the clocks.

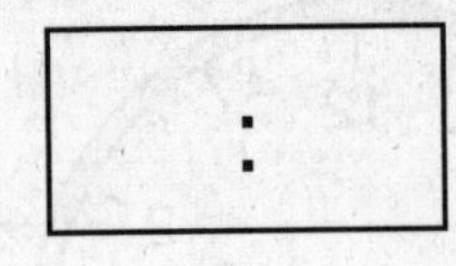

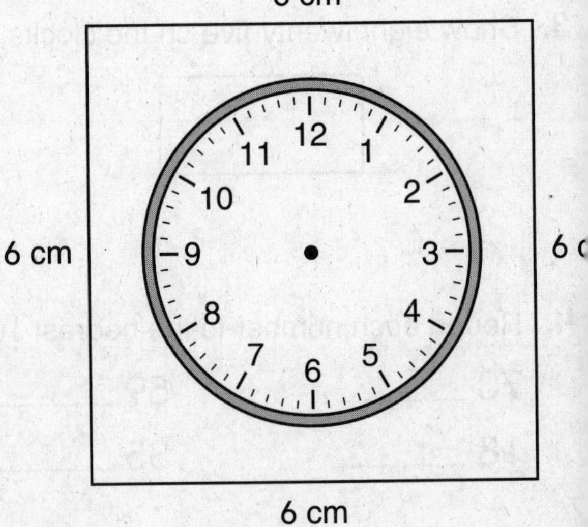

4. Round each number to the nearest 10.

82 ______ 55 ______

35 ______ 16 ______

5. What is the perimeter of the square around the clock?

Number sentence ____________________

Perimeter ______

6. Find the answers.

$$\begin{array}{r} 65 \\ +\ 38 \\ \hline \end{array} \qquad \begin{array}{r} 65 \\ -\ 38 \\ \hline \end{array} \qquad \begin{array}{r} 56 \\ +\ 29 \\ \hline \end{array} \qquad \begin{array}{r} 56 \\ -\ 29 \\ \hline \end{array}$$

Name ______________________ Score ________

Fact Assessment 20-2

***Saxon** Math 2 (for use with **Lesson 105-2**)*

S90: 90 Subtraction Facts

1	7 − 1	10 − 4	9 − 0	5 − 4	12 − 6	9 − 7	11 − 3	8 − 2	6 − 6	9 − 4
2	6 − 3	11 − 6	10 − 2	6 − 1	12 − 8	2 − 0	9 − 3	7 − 2	6 − 5	5 − 5
3	11 − 4	4 − 2	8 − 0	10 − 6	14 − 5	4 − 0	12 − 7	9 − 5	8 − 4	13 − 8
4	9 − 2	11 − 5	15 − 6	5 − 1	8 − 5	16 − 8	8 − 6	11 − 7	1 − 0	7 − 3
5	9 − 6	4 − 3	17 − 8	10 − 5	12 − 4	13 − 7	8 − 3	16 − 7	10 − 3	4 − 1
6	6 − 2	13 − 5	7 − 0	11 − 2	10 − 8	10 − 7	14 − 6	1 − 1	12 − 3	7 − 5
7	2 − 1	11 − 8	7 − 7	2 − 2	12 − 5	3 − 1	15 − 7	10 − 1	6 − 0	13 − 4
8	5 − 2	9 − 8	3 − 0	7 − 6	13 − 6	3 − 3	14 − 8	9 − 1	6 − 4	7 − 4
9	8 − 7	4 − 4	15 − 8	3 − 2	5 − 0	5 − 3	8 − 8	14 − 7	0 − 0	8 − 1

Name ______________________

Date ______________________

Written Assessment 20

***Saxon** Math 2 (for use with **Lesson 105-2**)*

1. Joshua had 42 baseball cards. He gave 15 cards to Dana. How many cards does he have now?

Workspace

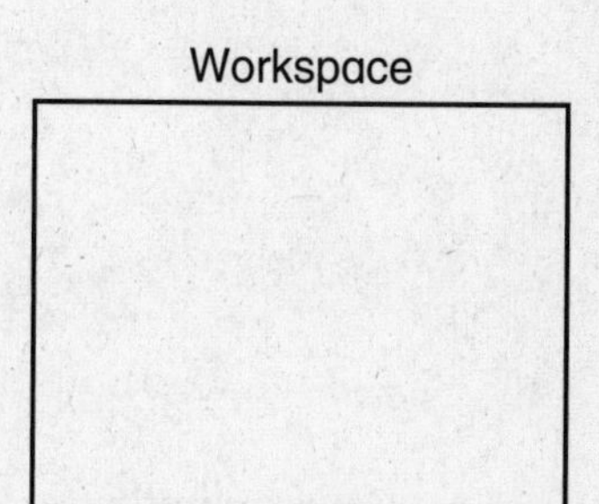

Number sentence ______________________

Answer ______________________

2. Draw a picture to show four hundred fifty-two. (Use ☐ for 100, ▯ for 10, and ▫ for 1.)

Write this number in expanded form. ______________________

Write four hundred fifty-two using digits. __________

3. Carla has 10 quarters. Draw the quarters. How much money is that? __________

4. Use the correct comparison symbol (>, <, or =).

2 + 8 ◯ 6 + 3   16 − 7 ◯ 10   20 + 30 ◯ 5 × 10

5. Find the products.

6 × 10 = ______   2 × 10 = ______   0 × 10 = ______

6. Find the differences.

$$\begin{array}{r} 80¢ \\ -\ 43¢ \\ \hline \end{array}$$

75¢ − 24¢

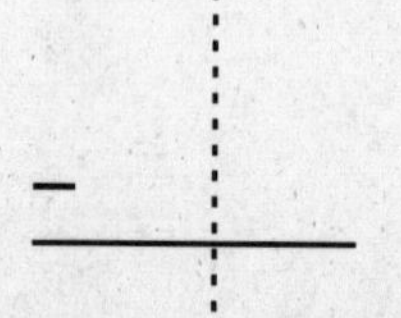

62¢ − 36¢

Name ____________________

Lesson Worksheet 105-2

Saxon *Math 2 (for use with* ***Lesson 105-2****)*

Our Favorite Ice Cream Flavors

1. Write two questions about the graph.

1. ____________________

2. ____________________

2. Write three facts about the graph.

1. ____________________

2. ____________________

3. ____________________

Name ____________________

Class Fact Practice 106A

***Saxon** Math 2 (for use with **Lesson 106**)*

Set 21: Subtracting 9 and 8; Review Facts

Do −9 Wrap-Up once.
Do −8 Wrap-Up once.

$15 - 9$	$11 - 9$	$13 - 9$	$17 - 9$	$9 - 9$
$14 - 9$	$18 - 9$	$10 - 9$	$16 - 9$	$12 - 9$
$16 - 8$	$11 - 8$	$15 - 8$	$13 - 8$	$8 - 8$
$10 - 8$	$17 - 8$	$14 - 8$	$9 - 8$	$12 - 8$
$7 - 7$	$16 - 7$	$14 - 7$	$9 - 7$	$10 - 7$

Name ______________________________

Saxon Math 2 (for use with Lesson 106)

Set 21: Subtracting 9 and 8; Review Facts

1. Read the answers to someone.
2. Write the answers.
3. Ask someone to correct your paper. Corrected by ____________________

12 − 9	17 − 9	14 − 9	9 − 9	13 − 9
16 − 9	18 − 9	11 − 9	15 − 9	10 − 9
10 − 8	16 − 8	12 − 8	8 − 8	15 − 8
13 − 8	17 − 8	9 − 8	14 − 8	11 − 8
16 − 7	10 − 7	7 − 7	11 − 7	14 − 7

Name ______________________________

Write the time.

:

Write the time.

:

Write the time.

:

Show the time.

3:29

Show the time.

7:08

Show the time.

12:54

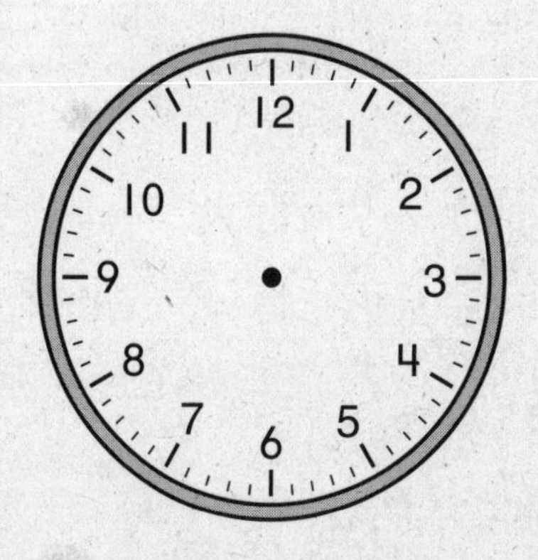

Name •
Draw an 8-cm line segment.

Guided Class Practice 106A

***Saxon** Math 2 (for use with **Lesson 106**)*

Date •————————————————•

Measure this line segment using centimeters. _______ cm

1. Ahmad put the peanuts in groups of 10. When he finished, he counted four groups of peanuts. Draw a picture and write a number sentence to show how many peanuts he has.

_______ groups of _______ peanuts. Number sentence _______________

How many peanuts does he have? _______

2. It's morning.

What time is it? _______________

3. Draw a line of symmetry in each shape.
Color one half of each shape.

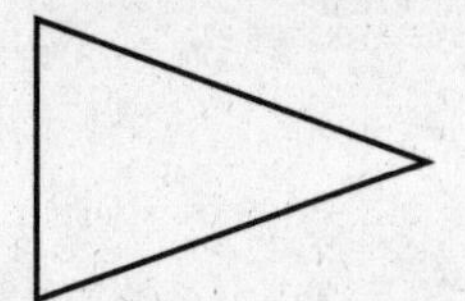

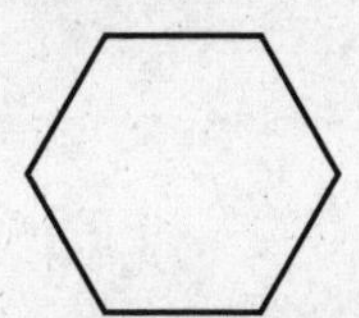

4. Mrs. Klisky's class is making toy cars. They will need 4 wheels for each car. Use the table to show the number of wheels they will need for 10 cars.

Cars	1	2	3	4	5	6				
Wheels	4	8	12							

How many wheels will 10 cars have? _______________

5. Find the answers.

$7 \times 100 =$ _______

$3 \times 10 =$ _______

$4 \times 1 =$ _______

$5 \times 100 =$ _______

$2 + 6 + 3 + 7 + 4 =$ _______

$87 - 39$

$$\begin{array}{r} 78 \\ 24 \\ +\ 32 \\ \hline \end{array}$$

Name ______________________________

Date ______________________________

Saxon Math 2 (for use with Lesson 106)

1. Justin put the paper clips in groups of 10. When he finished, he counted six groups of paper clips. Draw a picture and write a number sentence to show how many paper clips he has.

_______ groups of _______ paper clips. Number sentence _______________

How many paper clips does he have? _______

2. It's afternoon.

What time is it? _______________

3. Draw a line of symmetry in each shape.
Color one half of each shape.

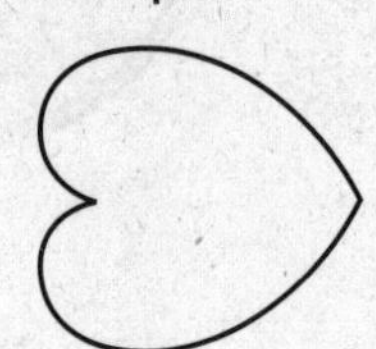

4. Mrs. Klisky's class is making toy cars. They will need 2 headlights for each car. Use the table to show the number of headlights they will need for 10 cars.

Cars	1	2	3	4	5	6				
Headlights	2	4	6							

How many headlights will 10 cars have? _______________

5. Find the answers.

$9 \times 1 =$ _______ $7 \times 10 =$ _______

$6 \times 100 =$ _______ $3 \times 100 =$ _______

$4 + 2 + 9 + 3 + 1 + 6 =$ _______

$73 - 29$

$$\begin{array}{r} 84 \\ 93 \\ +\ 16 \\ \hline \end{array}$$

Name ____________________

Class Fact Practice 107A

***Saxon** Math 2 (for use with **Lesson 107**)*

Set 21: Subtracting 9 and 8

Do –9 Wrap-Up twice. ☐☐ Do –8 Wrap-Up twice. ☐☐

11 – 9 = ____	8 – 8 = ____
9 – 9 = ____	16 – 8 = ____
14 – 9 = ____	12 – 8 = ____
18 – 9 = ____	9 – 8 = ____
10 – 9 = ____	15 – 8 = ____
15 – 9 = ____	13 – 8 = ____
12 – 9 = ____	10 – 8 = ____
17 – 9 = ____	17 – 8 = ____
13 – 9 = ____	14 – 8 = ____
16 – 9 = ____	11 – 8 = ____

Name ________________________________

Fact Homework 107B

***Saxon** Math 2 (for use with **Lesson 107**)*

Set 21: Subtracting 9 and 8

1. Write the answers.
2. Draw lines to connect the problems that have the same answer.
3. Ask someone to correct your paper. Corrected by ____________________

11 − 9 = ____ •	• 15 − 8 = ____
16 − 9 = ____ •	• 10 − 8 = ____
12 − 9 = ____ •	• 17 − 8 = ____
18 − 9 = ____ •	• 9 − 8 = ____
15 − 9 = ____ •	• 11 − 8 = ____
10 − 9 = ____ •	• 12 − 8 = ____
13 − 9 = ____ •	• 14 − 8 = ____
17 − 9 = ____ •	• 8 − 8 = ____
9 − 9 = ____ •	• 13 − 8 = ____
14 − 9 = ____ •	• 16 − 8 = ____

Name .

Draw a 7-cm line segment.

Date .____.

Measure this line segment using centimeters. ______ cm

1. The Grade 2 children at Rideout Elementary School collected cans for recycling. During the first week they collected 37 cans, during the second week they collected 88 cans, and during the third week they collected 96 cans. How many cans did they collect during the first two weeks?

Workspace

Number sentence ______________________

Answer ______________________

During which week were the most cans collected? __________

2. How much money is this? Write the amount two ways. __________ __________

______ "

______ "

3. Measure the sides of the rectangle in Problem 2 using inches. What is the perimeter?

Number sentence ____________________________ Perimeter ______

4. Show 5:47 on the clock.

5. Round each number to the nearest 10.

17 ______ 55 ______ 72 ______

6. Find the differences.

65 – 28

–

82 – 49

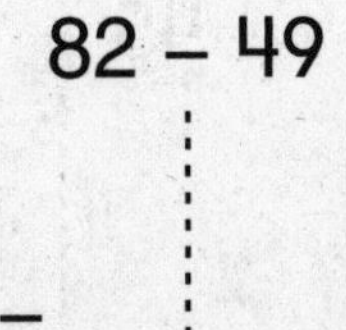

83 – 42

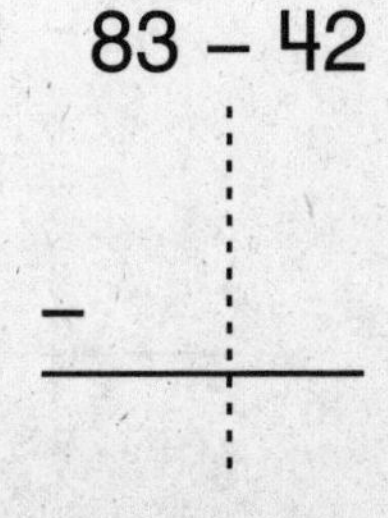

Name ______________________________

Date ______________________________

Saxon Math 2 (for use with **Lesson** 107)

1. The Grade 2 children in Mrs. Haller's class and Mrs. Carrell's class collected bottles for recycling. During the first week they collected 63 bottles, during the second week they collected 49 bottles, and during the third week they collected 32 bottles. How many bottles did they collect during the first two weeks?

Workspace

Number sentence ______________________________

Answer ______________________________

During which week were the most bottles collected? ____________

2. How much money is this? Write the amount two ways. ____________

3. Someone measured the sides of the rectangle in Problem 2. What is the perimeter?

Number sentence ______________________________ Perimeter ________

4. Show 10:16 on the clock.

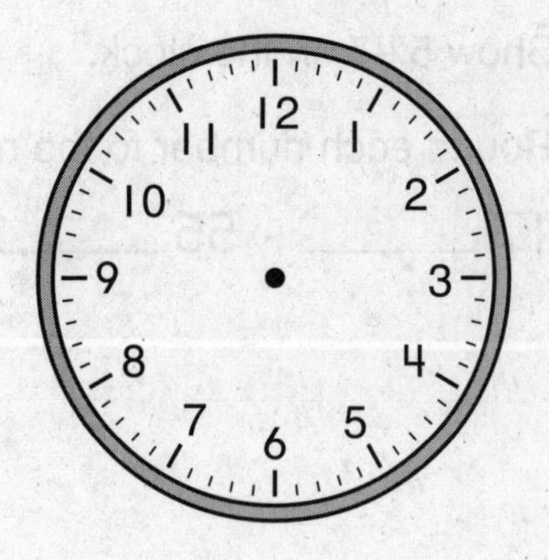

5. Round each number to the nearest 10.

54 ________ 38 ________ 65 ________

6. Find the differences.

72 – 53

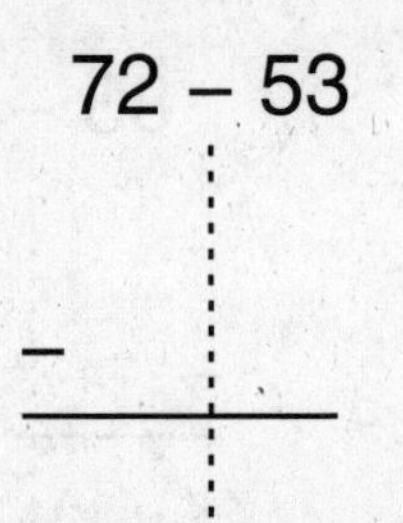

45 – 19

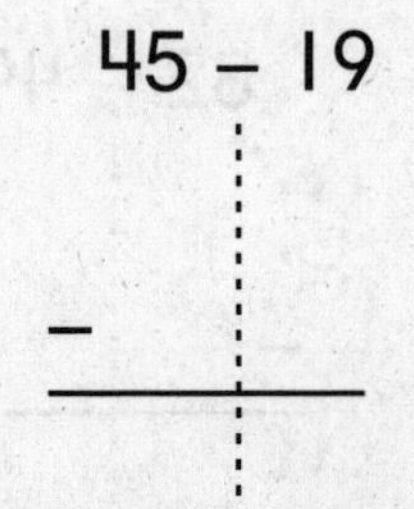

73 – 51

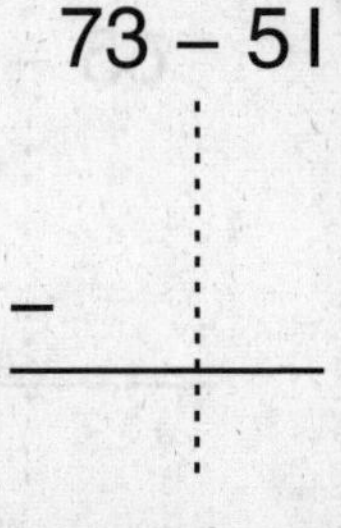

Name ____________________

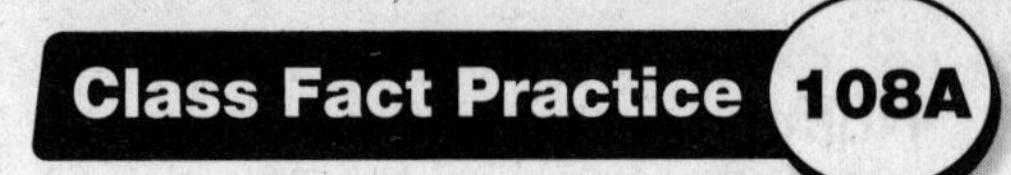

Saxon *Math 2 (for use with* ***Lesson 108****)*

Set 21: Subtracting 9 and 8

$14 - 9$	$9 - 8$	$17 - 9$	$9 - 9$	$11 - 8$
$16 - 9$	$14 - 8$	$10 - 9$	$12 - 8$	$15 - 9$
$13 - 8$	$18 - 9$	$15 - 8$	$11 - 9$	$17 - 8$
$15 - 9$	$12 - 9$	$10 - 8$	$16 - 9$	$14 - 9$
$8 - 8$	$17 - 9$	$13 - 9$	$16 - 8$	$18 - 9$

Name ______________________________

Saxon Math 2 (for use with ***Lesson 108****)*

Set 21: Subtracting 9 and 8

1. Read the answers to someone.
2. Write the answers.
3. Ask someone to correct your paper. Corrected by ____________________

$11 - 8$	$9 - 9$	$17 - 9$	$9 - 8$	$14 - 9$
$15 - 9$	$12 - 8$	$10 - 9$	$14 - 8$	$16 - 9$
$17 - 8$	$11 - 9$	$15 - 8$	$18 - 9$	$13 - 8$
$14 - 9$	$16 - 9$	$10 - 8$	$12 - 9$	$15 - 9$
$18 - 9$	$16 - 8$	$13 - 9$	$17 - 9$	$8 - 8$

Name

Draw an 8-cm line segment.

Date

Measure this line segment using centimeters. ________ cm

Guided Class Practice 108A

Saxon Math 2 (for use with Lesson 108)

1. There are 16 markers in a package. Kristina will share them equally with her cousin Michael.

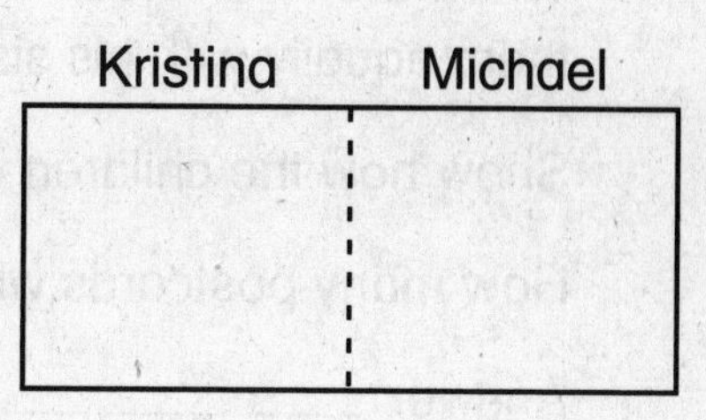

Show how the children will share the markers.

How many markers will each child have?

Answer ________________

2. Gina has 2 quarters, 3 dimes, 2 nickels, and a penny. Draw the coins.

How much money does she have?

Write the amount two ways. ________ ________

3. What is the perimeter of this shape?

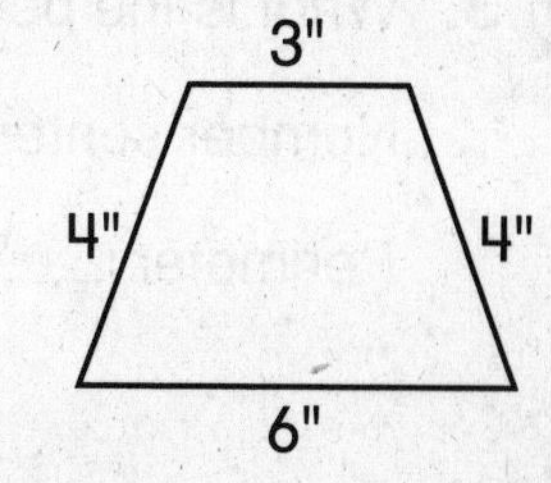

Number sentence ________________

Perimeter ________________

4. Use a red crayon to trace the parallel line segments in the shape in Problem 3.

5. Color the pyramid red.
Color the cylinder blue.
Color the sphere yellow.

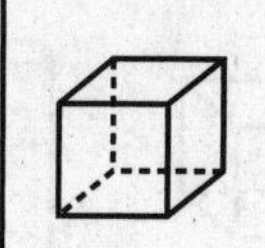 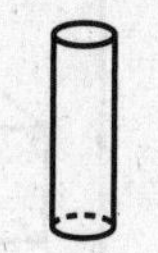 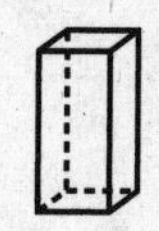 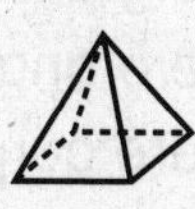  

6. Find the answers.

$$\begin{array}{r} 65 \\ +\ 49 \\ \hline \end{array} \qquad \begin{array}{r} 85¢ \\ +\ 9¢ \\ \hline \end{array} \qquad \begin{array}{r} 78 \\ -\ 39 \\ \hline \end{array} \qquad \begin{array}{r} 82 \\ -\ 39 \\ \hline \end{array}$$

Name ____________________

Date ____________________

Saxon Math 2 (for use with Lesson 108)

1. There are 10 postcards in a package. Bruce will share them equally with his sister Sarah.

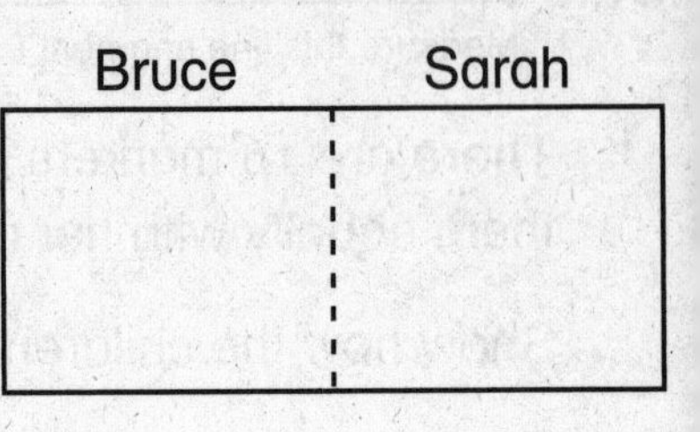

Show how the children will share the postcards.

How many postcards will each child have?

Answer ____________________

2. Daniel has 1 quarter, 2 dimes, 4 nickels, and 2 pennies. Draw the coins.

How much money does he have?

Write the amount two ways. __________ __________

3. What is the perimeter of this shape?

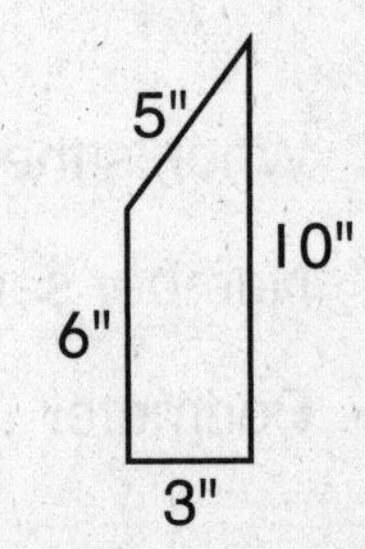

Number sentence ____________________

Perimeter ____________________

4. Use a red crayon to trace the parallel line segments in the shape in Problem 3.

5. Color the cone green.
Color the cube orange.
Color the sphere purple.

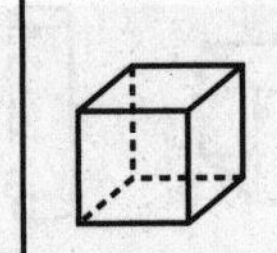
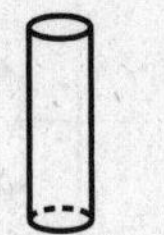
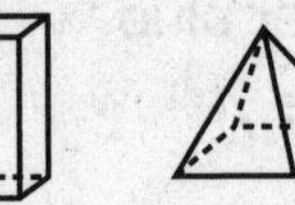
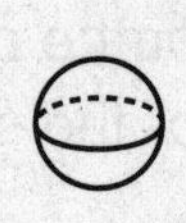

6. Find the answers.

$$\begin{array}{r} 74 \\ +\ 84 \\ \hline \end{array} \qquad \begin{array}{r} 64¢ \\ +\ 7¢ \\ \hline \end{array} \qquad \begin{array}{r} 78 \\ -\ 29 \\ \hline \end{array} \qquad \begin{array}{r} 52 \\ -\ 17 \\ \hline \end{array}$$

Name ______________________________

Class Fact Practice 109A

***Saxon** Math 2 (for use with **Lesson 109**)*

Set 21: Subtracting 9 and 8

Do –9 Wrap-Up twice.
Do –8 Wrap-Up once.

$10 - 9$ $\quad$ $11 - 8$ $\quad$ $18 - 9$ $\quad$ $8 - 8$ $\quad$ $13 - 9$

$15 - 8$ $\quad$ $9 - 9$ $\quad$ $17 - 8$ $\quad$ $14 - 9$ $\quad$ $9 - 8$

$15 - 9$ $\quad$ $12 - 8$ $\quad$ $16 - 8$ $\quad$ $11 - 9$ $\quad$ $13 - 8$

$14 - 8$ $\quad$ $12 - 9$ $\quad$ $17 - 9$ $\quad$ $10 - 8$ $\quad$ $16 - 9$

$10 - 9$ $\quad$ $16 - 8$ $\quad$ $14 - 8$ $\quad$ $18 - 9$ $\quad$ $13 - 9$

Saxon Math 2 (for use with **Lesson 109**)

Name ______________________________

Set 21: Subtracting 9 and 8

1. Read the answers to someone.
2. Write the answers.
3. Ask someone to correct your paper. Corrected by ____________________

11 − 9	15 − 8	18 − 9	15 − 9	10 − 8
13 − 8	12 − 9	8 − 8	16 − 8	13 − 9
16 − 9	14 − 8	11 − 8	9 − 9	17 − 9
12 − 8	10 − 9	17 − 8	14 − 9	9 − 8
14 − 8	18 − 9	10 − 9	16 − 8	13 − 9

Name. Draw a 7-cm line segment.

Date •———————————————————•

Measure this line segment using centimeters. ________ cm

Guided Class Practice 109A

Saxon Math 2 (for use with Lesson 109)

1. The children in Room 6 collected 16 quarters. They spent 9 quarters for new markers for the classroom. How many quarters do they have left?

Number sentence ________________________________

Answer ____________________ How much money is that? ____________

2. Show how two children will share the pattern blocks equally.

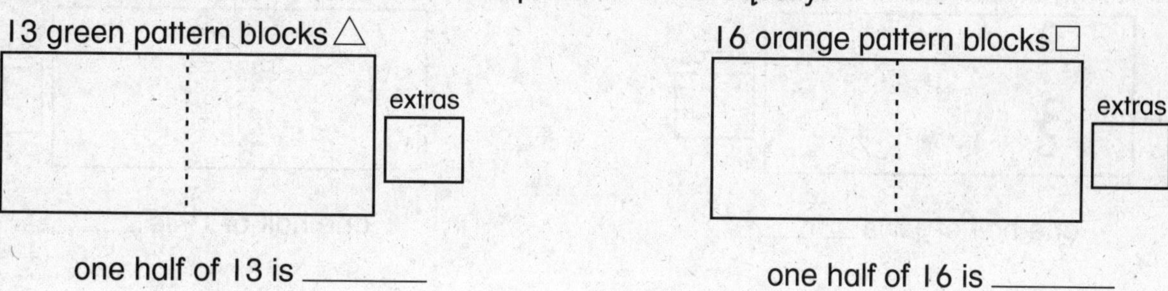

one half of 13 is ________ one half of 16 is ________

3. Circle the best number sentence to use to estimate the sum of 16 and 79.

$10 + 70 = 80$ $10 + 80 = 90$ $20 + 70 = 90$ $20 + 80 = 100$

4. Find the sums.

$$\begin{array}{r} 217 \\ +\ 347 \\ \hline \end{array} \qquad \begin{array}{r} 451 \\ +\ 196 \\ \hline \end{array} \qquad \begin{array}{r} \$1.75 \\ +\ \ 3.15 \\ \hline \end{array}$$

5. What is something you can do in 1 minute?

____

6. Find the answers.

$3 \times 10 =$ ________ $6 \times 100 =$ ________

$10 \times 10 =$ ________ $4 \times 1 =$ ________

$$\begin{array}{r} 86 \\ -\ 28 \\ \hline \end{array} \qquad \begin{array}{r} 98 \\ -\ 39 \\ \hline \end{array}$$

Name ______________________

Date ______________________

Saxon Math 2 (for use with Lesson 109)

1. Sharon had 4 quarters. Her sister gave her 2 more quarters. How many quarters does Sharon have now?

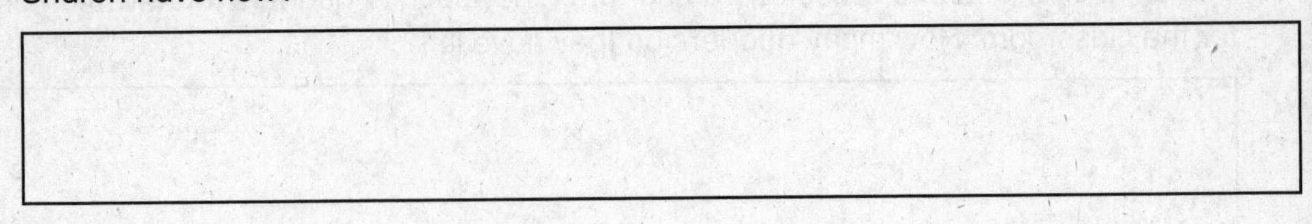

Number sentence ______________________

Answer ______________ How much money is that? ________

2. Show how two children will share the pattern blocks equally.

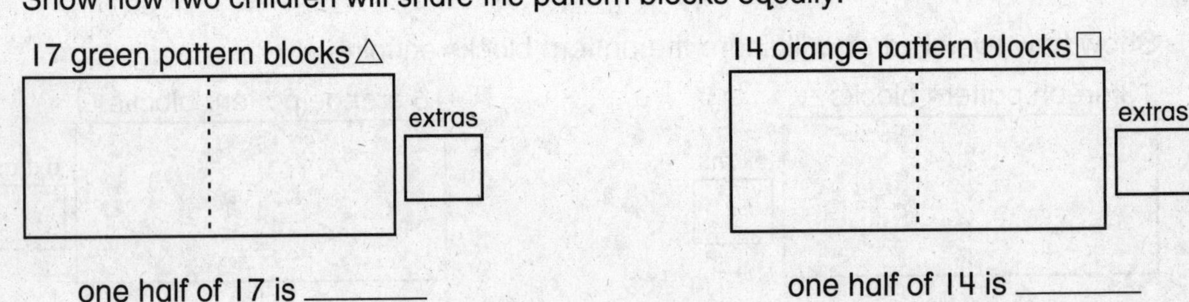

one half of 17 is ______ one half of 14 is ______

3. Circle the best number sentence to use to estimate the sum of 12 and 71.

$10 + 70 = 80$ $10 + 80 = 90$ $20 + 70 = 90$ $20 + 80 = 100$

4. Find the sums.

$$\begin{array}{r} 536 \\ +\ 246 \\ \hline \end{array} \qquad \begin{array}{r} 231 \\ +\ 284 \\ \hline \end{array} \qquad \begin{array}{r} \$4.27 \\ +\ \ 1.43 \\ \hline \end{array}$$

5. What is something you can do in 1 second?

____

6. Find the answers.

$5 \times 1 =$ ______ $7 \times 100 =$ ______

$9 \times 10 =$ ______ $10 \times 10 =$ ______

$$\begin{array}{r} 91 \\ -\ 35 \\ \hline \end{array} \qquad \begin{array}{r} 53 \\ -\ 19 \\ \hline \end{array}$$

Name ______________________ Score ________

Fact Assessment 21-1

***Saxon** Math 2 (for use with **Lesson 110-1**)*

Set 21: Subtracting 9 and 8

$14 - 9$	$11 - 8$	$10 - 9$	$8 - 8$	$13 - 9$
$15 - 8$	$12 - 9$	$16 - 9$	$12 - 8$	$11 - 9$
$14 - 8$	$17 - 9$	$9 - 9$	$13 - 8$	$15 - 9$
$9 - 8$	$11 - 9$	$17 - 8$	$10 - 9$	$12 - 9$
$18 - 9$	$10 - 8$	$13 - 9$	$16 - 8$	$9 - 9$

Name ______________________________

Fact Homework 110A

***Saxon** Math 2 (for use with **Lesson 110-1**)*

Set 21: Subtracting 9 and 8

Pretend you are the teacher.
Correct this paper.
If the answer is incorrect, write the correct answer next to the problem.

$$\begin{array}{r}14\\-\ 9\\\hline 5\end{array}\qquad\begin{array}{r}11\\-\ 8\\\hline 3\end{array}\qquad\begin{array}{r}10\\-\ 9\\\hline 1\end{array}\qquad\begin{array}{r}8\\-\ 8\\\hline 0\end{array}\qquad\begin{array}{r}13\\-\ 9\\\hline 5\end{array}$$

$$\begin{array}{r}15\\-\ 8\\\hline 7\end{array}\qquad\begin{array}{r}12\\-\ 9\\\hline 3\end{array}\qquad\begin{array}{r}16\\-\ 9\\\hline 7\end{array}\qquad\begin{array}{r}12\\-\ 8\\\hline 4\end{array}\qquad\begin{array}{r}11\\-\ 9\\\hline 2\end{array}$$

$$\begin{array}{r}14\\-\ 8\\\hline 7\end{array}\qquad\begin{array}{r}17\\-\ 9\\\hline 8\end{array}\qquad\begin{array}{r}9\\-\ 9\\\hline 0\end{array}\qquad\begin{array}{r}13\\-\ 8\\\hline 5\end{array}\qquad\begin{array}{r}15\\-\ 9\\\hline 7\end{array}$$

$$\begin{array}{r}9\\-\ 8\\\hline 1\end{array}\qquad\begin{array}{r}11\\-\ 9\\\hline 3\end{array}\qquad\begin{array}{r}17\\-\ 8\\\hline 9\end{array}\qquad\begin{array}{r}10\\-\ 9\\\hline 1\end{array}\qquad\begin{array}{r}12\\-\ 9\\\hline 3\end{array}$$

$$\begin{array}{r}18\\-\ 9\\\hline 9\end{array}\qquad\begin{array}{r}10\\-\ 8\\\hline 2\end{array}\qquad\begin{array}{r}13\\-\ 9\\\hline 4\end{array}\qquad\begin{array}{r}16\\-\ 8\\\hline 8\end{array}\qquad\begin{array}{r}9\\-\ 9\\\hline 0\end{array}$$

Name ____________________________

Saxon Math 2 (for use with **Lesson 110-1**)

S100: 100 Subtraction Facts Corrected by ________

1	7 − 1	10 − 4	9 − 0	16 − 9	5 − 4	12 − 6	9 − 7	11 − 3	8 − 2	6 − 6
2	9 − 4	6 − 3	11 − 6	10 − 2	6 − 1	12 − 8	2 − 0	9 − 3	7 − 2	6 − 5
3	5 − 5	11 − 4	4 − 2	15 − 9	8 − 0	10 − 6	14 − 5	9 − 9	4 − 0	12 − 7
4	9 − 5	17 − 9	8 − 4	13 − 8	9 − 2	11 − 5	15 − 6	5 − 1	8 − 5	16 − 8
5	8 − 6	11 − 7	1 − 0	7 − 3	9 − 6	4 − 3	17 − 8	10 − 5	12 − 4	13 − 7
6	8 − 3	16 − 7	10 − 3	4 − 1	6 − 2	13 − 5	7 − 0	14 − 9	11 − 2	10 − 8
7	13 − 9	10 − 7	18 − 9	14 − 6	1 − 1	12 − 3	7 − 5	2 − 1	11 − 8	7 − 7
8	2 − 2	12 − 5	3 − 1	15 − 7	10 − 1	6 − 0	13 − 4	5 − 2	9 − 8	3 − 0
9	11 − 9	7 − 6	13 − 6	3 − 3	14 − 8	9 − 1	6 − 4	12 − 9	7 − 4	8 − 7
10	4 − 4	15 − 8	3 − 2	5 − 0	5 − 3	8 − 8	14 − 7	10 − 9	0 − 0	8 − 1

Name ______________________________

Lesson Worksheet 110-1A

Saxon *Math 2 (for use with* ***Lesson 110-1****)*

A. 3×5 pennies = _____ pennies

5 pennies
$\times$ 3
___ pennies

B. 1×5 pennies = _____ pennies

5 pennies
$\times$ 1
___ pennies

C. 4×5 pennies = _____ pennies

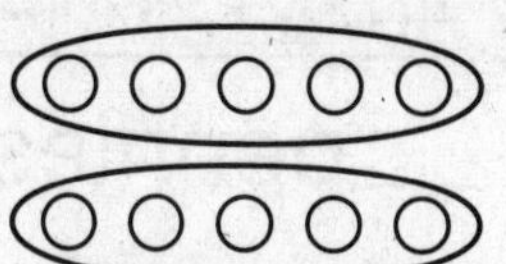

5 pennies
$\times$ 4
___ pennies

D. 2×5 pennies = _____ pennies

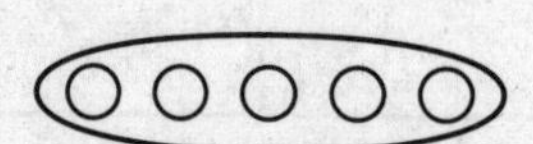

5 pennies
$\times$ 2
___ pennies

E. 0×5 pennies = _____ pennies

5 pennies
$\times$ 0
___ pennies

Name ______________________

Saxon Math 2 (for use with Lesson 110-1)

F. _____ × 5 pennies = _____ pennies

5 pennies
× ___
___ pennies

G. _____ × 5 pennies = _____ pennies

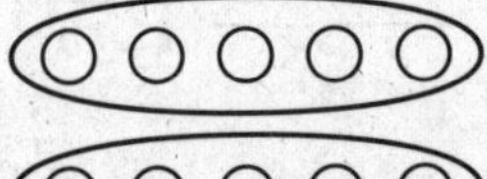

5 pennies
× ___
___ pennies

H. _____ × 5 pennies = _____ pennies

5 pennies
× ___

___ pennies

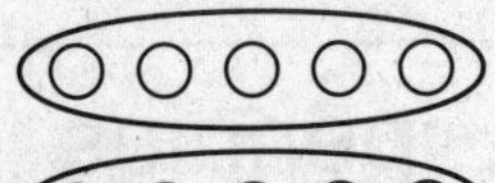

I. _____ × 5 pennies = _____ pennies

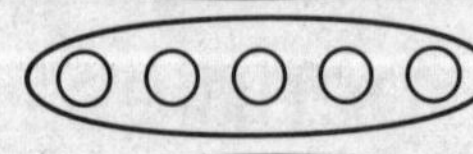

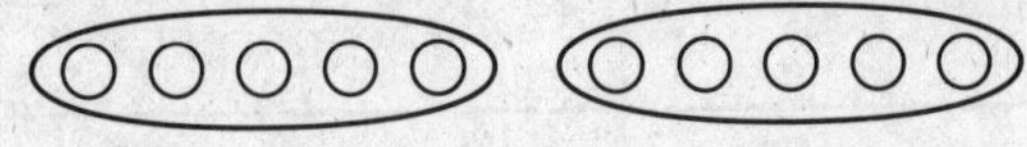

5 pennies
× ___
___ pennies

J. _____ × 5 pennies = _____ pennies

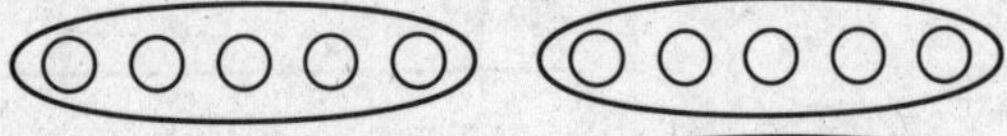

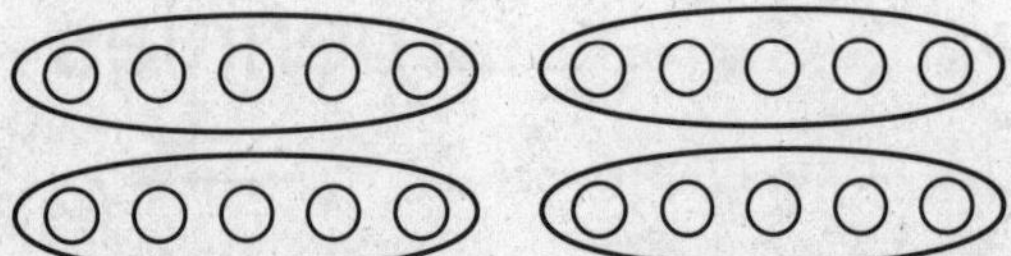

5 pennies
× ___
___ pennies

Name ______________________

Lesson Worksheet 110-1C

***Saxon** Math 2 (for use with **Lesson 110-1**)*

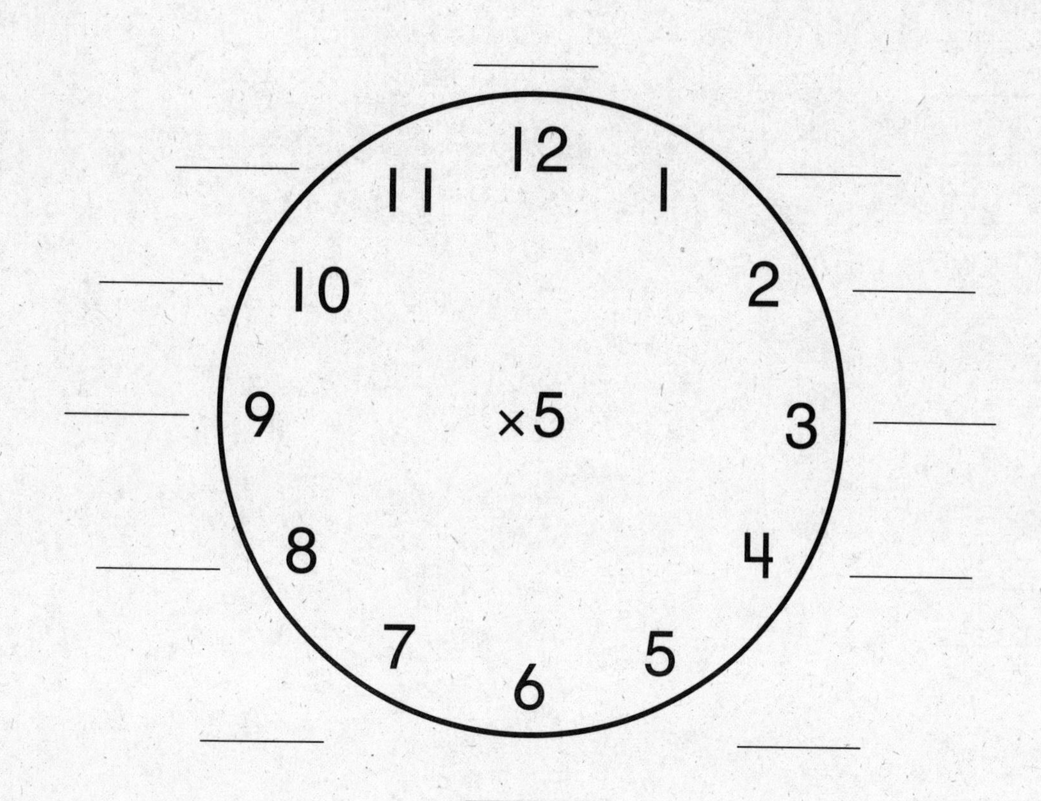

$1 \times 5 =$ ______ $5 \times 5 =$ ______ $9 \times 5 =$ ______

$2 \times 5 =$ ______ $6 \times 5 =$ ______ $10 \times 5 =$ ______

$3 \times 5 =$ ______ $7 \times 5 =$ ______ $11 \times 5 =$ ______

$4 \times 5 =$ ______ $8 \times 5 =$ ______ $12 \times 5 =$ ______

$\begin{array}{r} 5 \\ \times\ 0 \\ \hline \end{array}$ $\begin{array}{r} 5 \\ \times\ 1 \\ \hline \end{array}$ $\begin{array}{r} 5 \\ \times\ 2 \\ \hline \end{array}$ $\begin{array}{r} 5 \\ \times\ 3 \\ \hline \end{array}$ $\begin{array}{r} 5 \\ \times\ 4 \\ \hline \end{array}$ $\begin{array}{r} 5 \\ \times\ 5 \\ \hline \end{array}$ $\begin{array}{r} 5 \\ \times\ 6 \\ \hline \end{array}$ $\begin{array}{r} 5 \\ \times\ 7 \\ \hline \end{array}$ $\begin{array}{r} 5 \\ \times\ 8 \\ \hline \end{array}$ $\begin{array}{r} 5 \\ \times\ 9 \\ \hline \end{array}$

Name ______________________

Date ______________________

Problem-Solving Worksheet 110A

***Saxon** Math 2 (for use with **Lesson 110-1**)*

Understand	Plan	Solve	Check

Make an Organized List 

On Wednesdays the children at Westcliff School can buy hot dogs, hamburgers, or chicken nuggets. They can drink milk or juice. Show the different ways Kerry can buy one of the meals and one of the drinks.

Food ______________________
Drink ______________________

Food ______________________
Drink ______________________

Food ______________________
Drink ______________________

Food ______________________
Drink ______________________

Food ______________________
Drink ______________________

Food ______________________
Drink ______________________

Name ______________________

Saxon Math 2 (for use with Lesson 110-1)

Understand	Plan	Solve	Check

Mr. Bribiesca has orange, blue, and green folders for the children in his class. He has yellow and red pencils. Each child in his class can choose a folder and a pencil. Show the different ways Robert could choose a folder and a pencil.

____________ folder ____________ pencil	____________ folder ____________ pencil
____________ folder ____________ pencil	____________ folder ____________ pencil
____________ folder ____________ pencil	____________ folder ____________ pencil

Circle the problem-solving strategies you used to solve this problem.

Act It Out

Draw a Picture

Make an Organized List

Make a Table

Use Logical Reasoning

Look for a Pattern

Guess and Check

Explain how you got your answer: ______________________

____

____

____

Name ______________________ Score ________

Fact Assessment 21-2

Saxon *Math 2 (for use with* ***Lesson 110-2****)*

S100: 100 Subtraction Facts

1	7 − 1	10 − 4	9 − 0	16 − 9	5 − 4	12 − 6	9 − 7	11 − 3	8 − 2	6 − 6
2	9 − 4	6 − 3	11 − 6	10 − 2	6 − 1	12 − 8	2 − 0	9 − 3	7 − 2	6 − 5
3	5 − 5	11 − 4	4 − 2	15 − 9	8 − 0	10 − 6	14 − 5	9 − 9	4 − 0	12 − 7
4	9 − 5	17 − 9	8 − 4	13 − 8	9 − 2	11 − 5	15 − 6	5 − 1	8 − 5	16 − 8
5	8 − 6	11 − 7	1 − 0	7 − 3	9 − 6	4 − 3	17 − 8	10 − 5	12 − 4	13 − 7
6	8 − 3	16 − 7	10 − 3	4 − 1	6 − 2	13 − 5	7 − 0	14 − 9	11 − 2	10 − 8
7	13 − 9	10 − 7	18 − 9	14 − 6	1 − 1	12 − 3	7 − 5	2 − 1	11 − 8	7 − 7
8	2 − 2	12 − 5	3 − 1	15 − 7	10 − 1	6 − 0	13 − 4	5 − 2	9 − 8	3 − 0
9	11 − 9	7 − 6	13 − 6	3 − 3	14 − 8	9 − 1	6 − 4	12 − 9	7 − 4	8 − 7
10	4 − 4	15 − 8	3 − 2	5 − 0	5 − 3	8 − 8	14 − 7	10 − 9	0 − 0	8 − 1

Name ______________________________

Date ______________________________

Saxon Math 2 (for use with Lesson 110-2)

1. There are 22 children in Room 7. Twelve of these children are wearing sneakers. There are 24 children in Room 8. Fifteen of these children are wearing sneakers. Altogether, how many children are wearing sneakers?

Workspace

Number sentence ______________________________

Answer ______________________________

2. Glenn has 14 baseball cards. Show how he will share them equally with his sister.

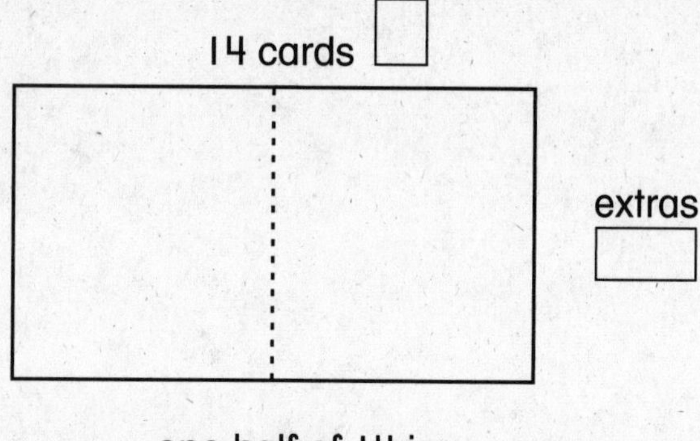

one half of 14 is ______

How many baseball cards will each child have? ______

3. Round each number to the nearest 10.

23 ______ 35 ______ 87 ______

4. Color the cone yellow.
Color the sphere red.
Color the cylinder blue.
Color the cube green.

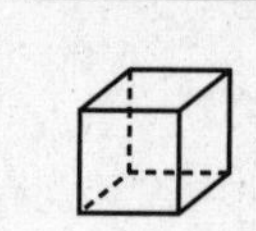   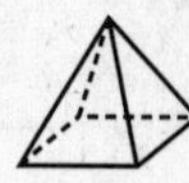  

5. Find the products.

$3 \times 100 =$ ______ $5 \times 10 =$ ______ $2 \times 1 =$ ______

6. Find the answers.

$$\begin{array}{r} 78 \\ +\ 37 \\ \hline \end{array} \qquad \begin{array}{r} 39 \\ 16 \\ +\ 83 \\ \hline \end{array} \qquad \begin{array}{r} 72 \\ -\ 27 \\ \hline \end{array} \qquad \begin{array}{r} 96 \\ -\ 38 \\ \hline \end{array}$$

Name ______________________________

Class Fact Practice 111A

***Saxon** Math 2 (for use with **Lesson 111**)*

Set 22: Multiplying by 5

5×0 | 5×1 | 5×2 | 5×3 | 5×4

5×5 | 5×6 | 5×7 | 5×8 | 5×9

5×6 | 5×2 | 5×8 | 5×0 | 5×4

5×5 | 5×1 | 5×9 | 5×7 | 5×3

5×4 | 5×6 | 5×5 | 5×8 | 5×9

Name ______________________________

Saxon Math 2 (for use with **Lesson 111**)

Set 22: Multiplying by 5

1. Read the answers to someone.
2. Write the answers.
3. Ask someone to correct your paper. Corrected by ______________________

$5 \times 0 =$ ____ $5 \times 1 =$ ____ $5 \times 2 =$ ____ $5 \times 3 =$ ____ $5 \times 4 =$ ____

$5 \times 5 =$ ____ $5 \times 6 =$ ____ $5 \times 7 =$ ____ $5 \times 8 =$ ____ $5 \times 9 =$ ____

$5 \times 4 =$ ____ $5 \times 1 =$ ____ $5 \times 8 =$ ____ $5 \times 2 =$ ____ $5 \times 5 =$ ____

$5 \times 3 =$ ____ $5 \times 9 =$ ____ $5 \times 0 =$ ____ $5 \times 6 =$ ____ $5 \times 7 =$ ____

$5 \times 8 =$ ____ $5 \times 4 =$ ____ $5 \times 6 =$ ____ $5 \times 3 =$ ____ $5 \times 9 =$ ____

Name ______________________________

⬡ = 1 Write a mixed number to show the value of the pattern blocks in each problem.

1. mixed number

2. mixed number

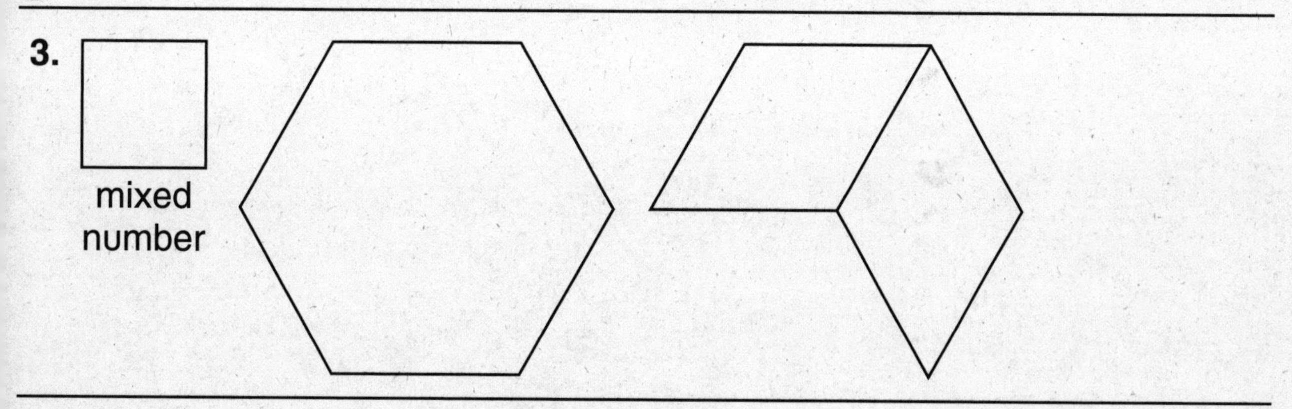

3. mixed number

Trace pattern blocks to show these mixed numbers.

4. $1\frac{1}{3}$

5. Show $2\frac{3}{6}$ on the back of this paper.

Name •

Draw a 7-cm line segment.

Date •————————•

Measure this line segment using centimeters. ________ cm

Workspace

1. There are 247 children in Grade 1, 291 children in Grade 2, and 308 children in Grade 3.

Which grade has the most children? ____________

Which grade has the fewest children? ____________

How many children are there in Grades 1 and 2 altogether?

Number sentence ______________________________

Answer ______________________

2. I have 1 quarter, 2 dimes, 3 nickels, and 1 penny. Draw the coins.

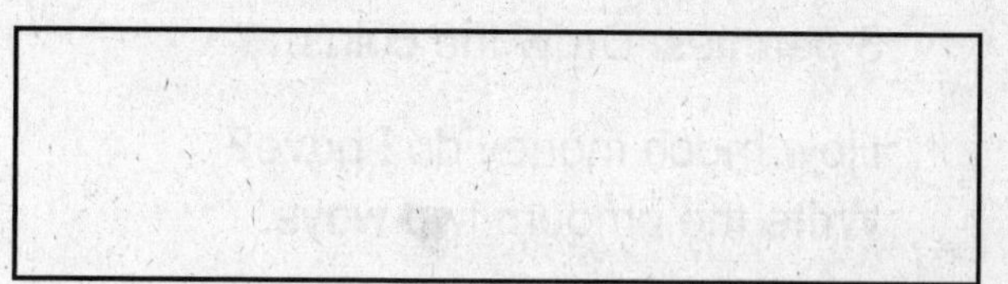

How much money do I have?
Write the amount two ways.

________ ________

3. Measure each side of the rectangle in Problem 2 using centimeters. Find the perimeter.

Number sentence ______________________________

What is the perimeter? ____________

4. Show seven twenty-nine on both clocks.

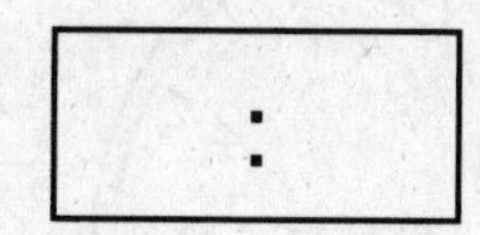

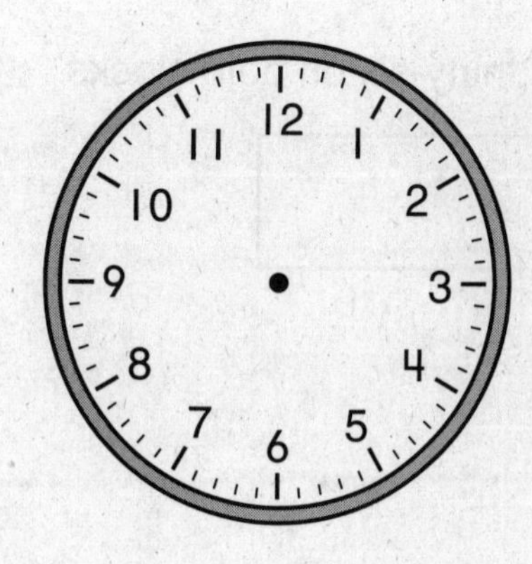

5. Which unit of measure is best for telling how much a dog would weigh?

ounce quart foot pound

6. Find the answers.

$$\begin{array}{r} 272 \\ +\ 318 \\ \hline \end{array} \qquad \begin{array}{r} \$1.28 \\ +\ 2.94 \\ \hline \end{array} \qquad \begin{array}{r} 75 \\ -\ 36 \\ \hline \end{array} \qquad \begin{array}{r} 46 \\ -\ 23 \\ \hline \end{array}$$

Name ____________________

Date ____________________

***Saxon** Math 2 (for use with **Lesson 111**)*

1. There are 195 children in Grade 4, 215 children in Grade 5, and 178 children in Grade 6.

Workspace

Which grade has the most children? ____________

Which grade has the fewest children? ____________

How many children are there in Grades 5 and 6 altogether?

Number sentence ____________________

Answer ____________________

2. I have 1 quarter, 2 dimes, 1 nickel, and 3 pennies. Draw the coins.

How much money do I have? Write the amount two ways.

__________ __________

7 cm

3 cm

3. Someone measured each side of the rectangle in Problem 2 using centimeters. Find the perimeter.

Number sentence ____________________

What is the perimeter? ____________

4. Show nine thirty-six on both clocks.

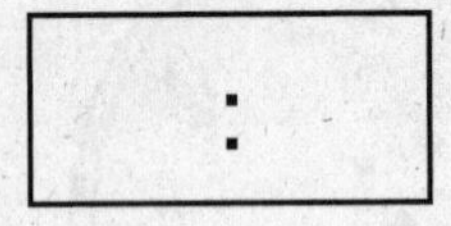

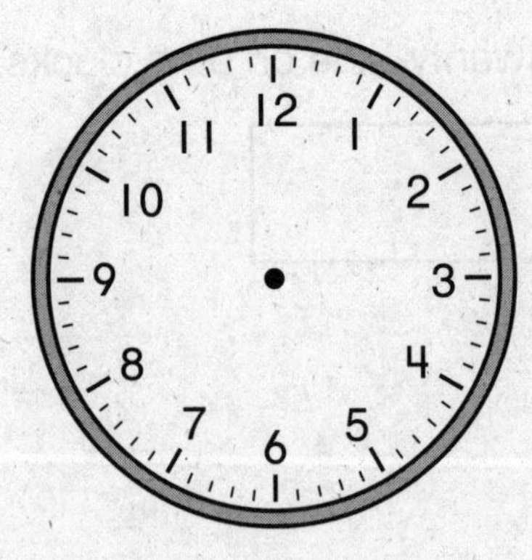

5. Which unit of measure is best for telling how much a child would weigh?

pint ounce pound inch

6. Find the answers.

$$\begin{array}{r} 327 \\ +\ 547 \\ \hline \end{array} \qquad \begin{array}{r} \$4.55 \\ +\ 2.73 \\ \hline \end{array} \qquad \begin{array}{r} 95 \\ -\ 27 \\ \hline \end{array} \qquad \begin{array}{r} 83 \\ -\ 62 \\ \hline \end{array}$$

Name ______________________________

Class Fact Practice 112A

***Saxon** Math 2 (for use with **Lesson 112**)*

Set 22: Multiplying by 5

1. Multiply each number on the clockface by 5.

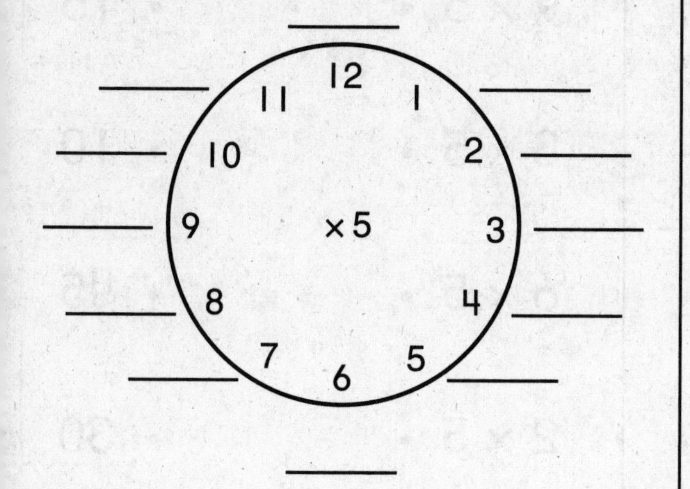

2. Match the problems to the answers.

2 × 5 •	• 15
6 × 5 •	• 10
3 × 5 •	• 45
9 × 5 •	• 30
7 × 5 •	• 20
0 × 5 •	• 35
4 × 5 •	• 40
8 × 5 •	• 25
1 × 5 •	• 0
5 × 5 •	• 5

3. Fill in the missing factors.

☐ × 5 = 35

☐ × 5 = 20

☐ × 5 = 30

☐ × 5 = 45

☐ × 5 = 15

Name ____________________

Saxon Math 2 (for use with **Lesson 112**)

Set 22: Multiplying by 5 Corrected by ________

1. Multiply each number on the clockface by 5.

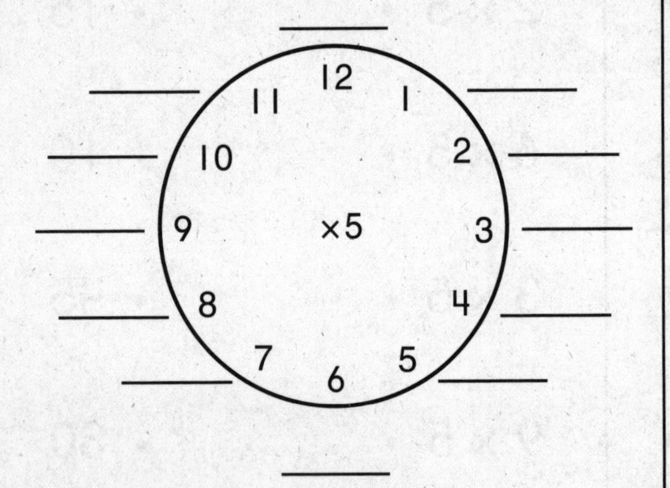

2. Match the problems to the answers.

9 × 5 •	• 15
3 × 5 •	• 10
6 × 5 •	• 45
2 × 5 •	• 30
8 × 5 •	• 20
4 × 5 •	• 35
7 × 5 •	• 40
1 × 5 •	• 25
5 × 5 •	• 0
0 × 5 •	• 5

3. Fill in the missing factors.

☐ × 5 = 10

☐ × 5 = 40

☐ × 5 = 25

☐ × 5 = 0

☐ × 5 = 35

Name ______________________________

Lesson Worksheet 112

***Saxon** Math 2 (for use with **Lesson 112**)*

1. Color $2\frac{1}{2}$ squares.

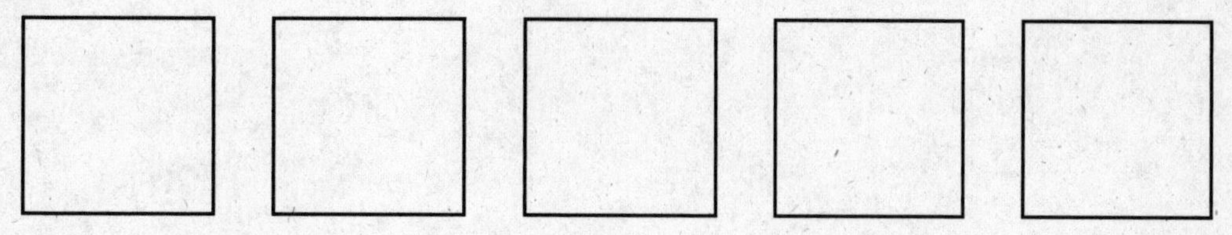

2. Color $3\frac{1}{2}$ circles.

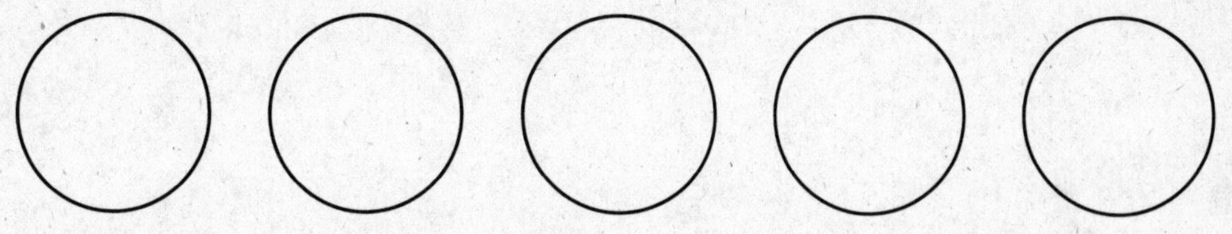

3. Color $4\frac{1}{4}$ squares.

4. Color $1\frac{3}{4}$ circles.

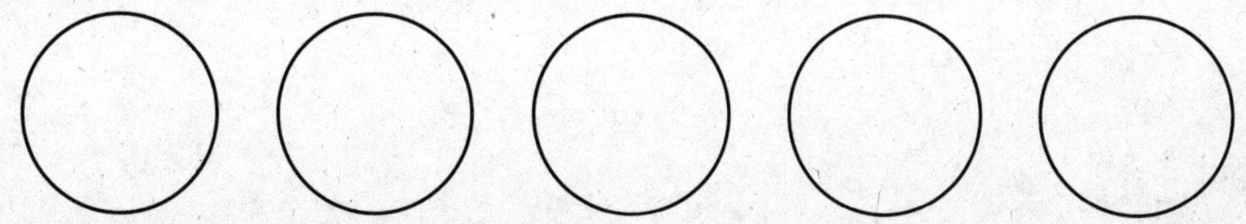

Write a mixed number to show how much is shaded.

5.

mixed number

6.

mixed number

7.

mixed number

8.

mixed number

Name

Draw an 8-cm line segment.

Date

Measure this line segment using centimeters. ______ cm

Workspace

1. The children in Mrs. Conte's class planted 18 tomato plants and 16 pepper plants. The children in Mrs. Mancano's class planted 14 tomato plants and 15 squash plants. How many tomato plants did they plant altogether?

Number sentence ______________________

Answer ______________________

2. How much money is this?
Write the amount two ways.

____

____

3. Measure the sides of the rectangle in Problem 2 using centimeters. What is the perimeter?

Number sentence ______________________

Perimeter ______________________

4. Write a mixed number to show how many squares are shaded.

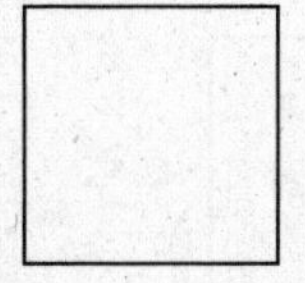

 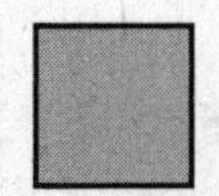  

5. How many pets do the children in Room 6 have altogether? ______

How many more dogs are there than birds? ______

Which two pets make up half of all the children's pets?

dogs and cats   birds and dogs   cats and birds

Room 6 Children's Pets

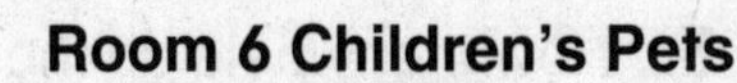

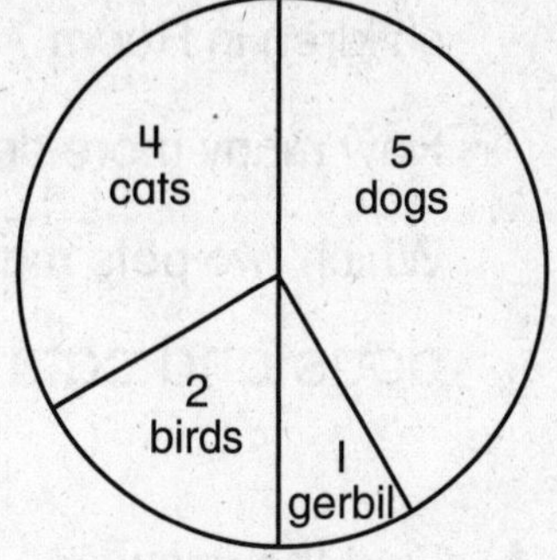

6. Find the answers.

$$\begin{array}{r} 82 \\ -\ 28 \\ \hline \end{array} \qquad \begin{array}{r} 346 \\ +\ 372 \\ \hline \end{array} \qquad \begin{array}{r} \$1.39 \\ +\ 5.68 \\ \hline \end{array}$$

$8 \times 5 =$ ______

$3 \times 5 =$ ______

Name ______________________________

Date ______________________________

Homework 112

Saxon Math 2 (for use with Lesson 112)

Workspace

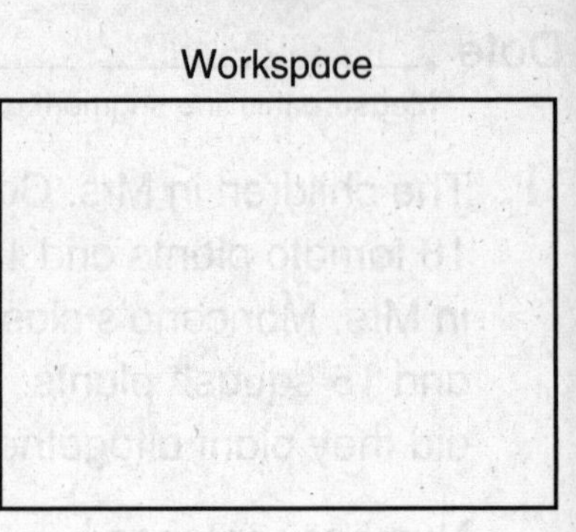

1. The children in Mrs. Ley's class planted 24 geraniums and 36 marigolds. The children in Mrs. Delmonte's class planted 38 geraniums and 20 marigolds. How many geraniums did the children plant altogether?

Number sentence ______________________________

Answer ______________________________

2. How much money is this? Write the amount two ways. __________ __________

11 cm

3 cm

3 cm

11 cm

3. Find the perimeter of the rectangle in Problem 2.

Number sentence ______________________________

Perimeter ______________________________

4. Write a mixed number to show how many circles are shaded.

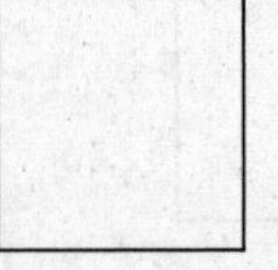

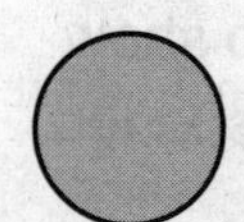

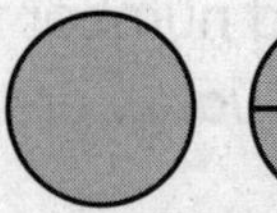

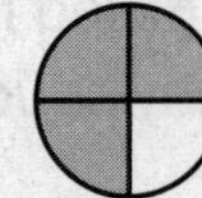

5. How many pets do the children in Room 7 have altogether? __________

How many more dogs are there than birds? __________

Which two pets make up half of all the children's pets?

dogs and cats   birds and dogs   cats and birds

Room 7 Children's Pets

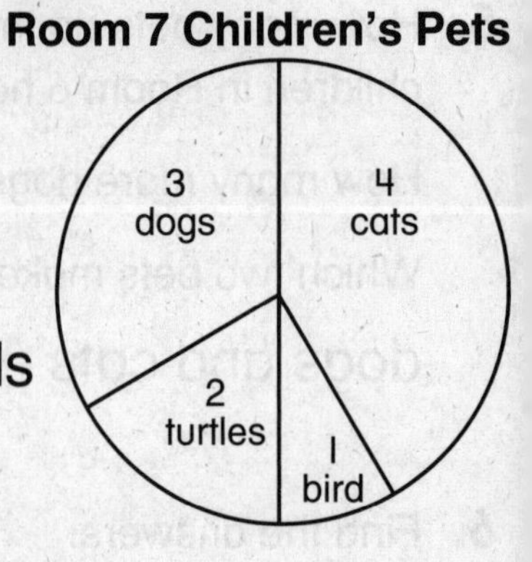

6. Find the answers.

$$\begin{array}{r} 73 \\ -\ 37 \\ \hline \end{array} \qquad \begin{array}{r} 163 \\ +\ 452 \\ \hline \end{array} \qquad \begin{array}{r} \$6.47 \\ +\ \ 2.57 \\ \hline \end{array}$$

$7 \times 5 =$ __________

$4 \times 5 =$ __________

Name ______________________________

Class Fact Practice **113A**

***Saxon** Math 2 (for use with **Lesson 113**)*

Set 22: Multiplying by 5

5×2	5×5	5×8	5×1	5×6
5×3	5×9	5×0	5×7	5×4
5×8	5×3	5×2	5×5	5×0
5×6	5×1	5×9	5×4	5×7
5×3	5×5	5×9	5×6	5×2

Name ____________________

Set 22: Multiplying by 5

1. Read the answers to someone.
2. Write the answers.
3. Ask someone to correct your paper. Corrected by ____________________

$5 \times 6 =$ ___ $5 \times 2 =$ ___ $5 \times 5 =$ ___ $5 \times 0 =$ ___ $5 \times 9 =$ ___

$5 \times 4 =$ ___ $5 \times 1 =$ ___ $5 \times 7 =$ ___ $5 \times 2 =$ ___ $5 \times 8 =$ ___

$5 \times 5 =$ ___ $5 \times 4 =$ ___ $5 \times 0 =$ ___ $5 \times 7 =$ ___ $5 \times 3 =$ ___

$5 \times 1 =$ ___ $5 \times 9 =$ ___ $5 \times 2 =$ ___ $5 \times 6 =$ ___ $5 \times 8 =$ ___

$5 \times 4 =$ ___ $5 \times 6 =$ ___ $5 \times 3 =$ ___ $5 \times 9 =$ ___ $5 \times 3 =$ ___

Name ______________________________

Mint Date of Pennies

Before 1980	1980–1984	1985–1989	1990–1994	1995–1999	2000–2004	2005–Now

	Before 1980	1980–1984	1985–1989	1990–1994	1995–1999	2000–2004	2005–Now
20							
18							
16							
14							
12							
10							
8							
6							
4							
2							
0							

How many pennies were minted in the years 1985 through 1989? ________

How many pennies were minted after 1989? ________

Which column has the most pennies? ____________________

Name •
Draw a 7-cm line segment.

Date •————————•
Measure this line segment using centimeters. ______ cm

1. The children in Mrs. Affinito's class chose their favorite fruits. Eight children chose bananas, fifteen children chose oranges, and seven children chose apples. Color the graph to show how many children chose each type of fruit.

Write a question for the graph. ______________________________

____

____

2. About how much might the lunch box of a child in your classroom weigh?

200 pounds   4 pounds   100 pounds   40 pounds

3. Measure each side of this shape using centimeters.

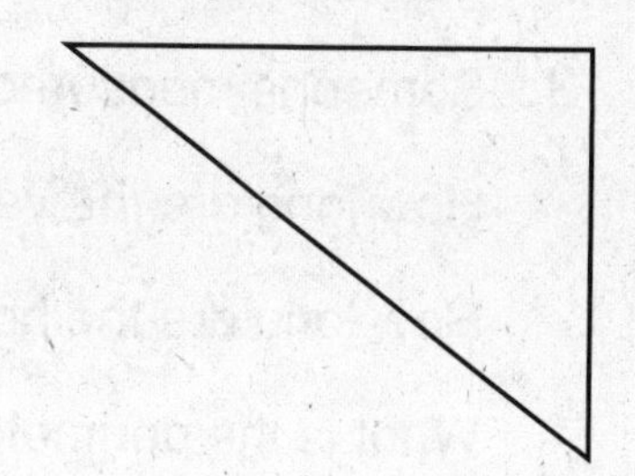

How long is the vertical line segment? ______ cm

How long is the oblique line segment? ______ cm

How long is the horizontal line segment? ______ cm

What is the perimeter? ______________________________

4. It's dark outside.

What time is it? ______________

5. Use the correct comparison symbol (>, <, or =).

35 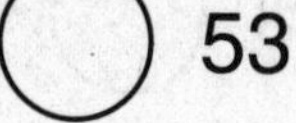 53   8 + 37  9 × 5   16 − 7 ◯ 14 − 6

6. Find the answers.

59 + 87   74 − 28   291 + 487

Name ____________________

Date ____________________

Saxon Math 2 (for use with **Lesson 113**)

1. The children in Mrs. Ramsey's class chose their favorite fruits. Six children chose bananas, thirteen children chose oranges, and nine children chose apples. Color the graph to show how many children chose each type of fruit.

Write a question for the graph. ____________________

____

____

2. About how much would a gallon of milk weigh?

100 pounds   50 pounds   8 pounds   1 pound

3. Someone measured each side of this shape using centimeters.

2 cm

4 cm

How long are the vertical line segments? ______ cm

How long are the horizontal line segments? ______ cm

What is the perimeter? ____________________

4. It's light outside.

What time is it? ____________________

5. Use the correct comparison symbol (>, <, or =).

68  86   3 + 28  6 × 5   17 − 8 ◯ 13 − 4

______ ______   ______ ______

6. Find the answers.

65 + 86   85 − 27   384 + 273

+ ______   − ______   + ______

Name ____________________

Class Fact Practice 114A

Saxon *Math 2 (for use with* ***Lesson 114****)*

Set 22: Multiplying by 5

5×3	5×9	5×4	5×1	5×6
5×8	5×0	5×5	5×7	5×2
5×6	5×4	5×9	5×1	5×8
5×3	5×7	5×2	5×5	5×0
5×7	5×1	5×4	5×8	5×3

Name ____________________

Saxon Math 2 (for use with Lesson 114)

Set 22: Multiplying by 5

1. Read the answers to someone.
2. Write the answers.
3. Ask someone to correct your paper. Corrected by ____________________

5×7	5×0	5×5	5×9	5×1
5×4	5×8	5×2	5×6	5×3
5×6	5×4	5×1	5×8	5×2
5×0	5×7	5×5	5×3	5×9
5×4	5×8	5×2	5×7	5×3

Name ______________________________

Shapes With Right Angles

1.

2.

3.

4.

Name •

Draw an 8-cm line segment.

Guided Class Practice 114A

***Saxon** Math 2 (for use with **Lesson 114**)*

Date •————————————————————•

Measure this line segment using centimeters. ________ cm

1. There are 43 jump ropes and 12 balls for the children to use during recess. Fifteen children are using jump ropes. How many jump ropes are not being used?

Number sentence ______________________

Answer ______________________

Workspace

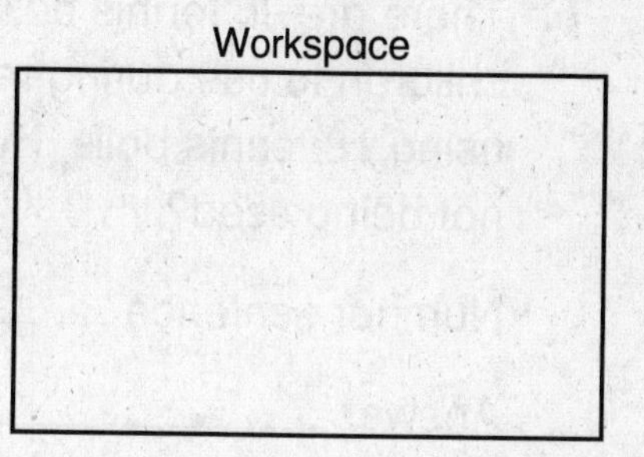

2. This is a tally to show how many children chose each color. Color the graph to show the colors the children chose.

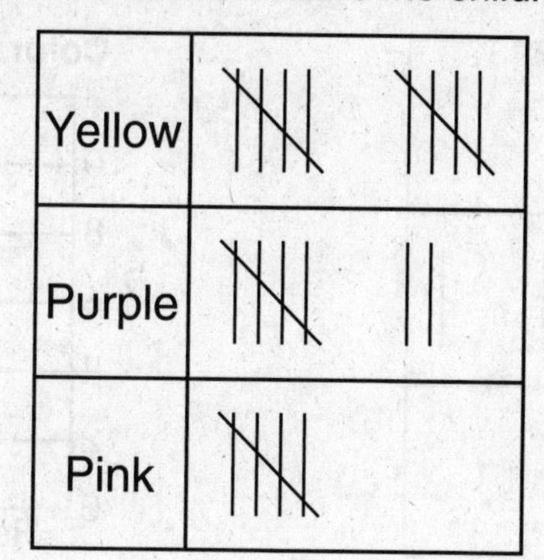

Colors Children Chose

12
10
8
6
4
2
0
Yellow Purple Pink

3. Color $4\frac{1}{4}$ squares.

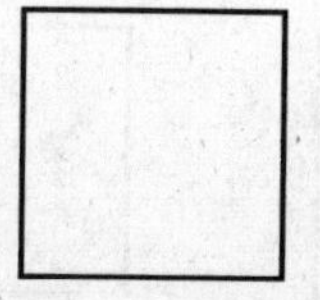 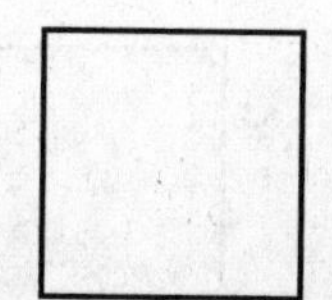 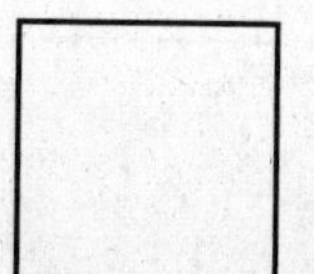 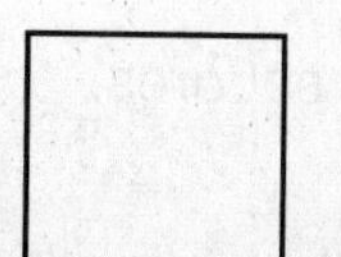 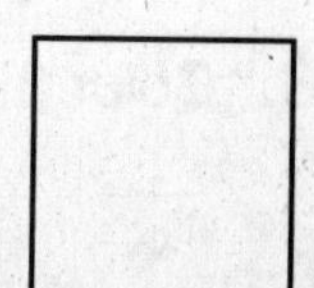

4. Put a small square in each right angle of this shape. (Use a corner of a piece of paper to check the angles.)

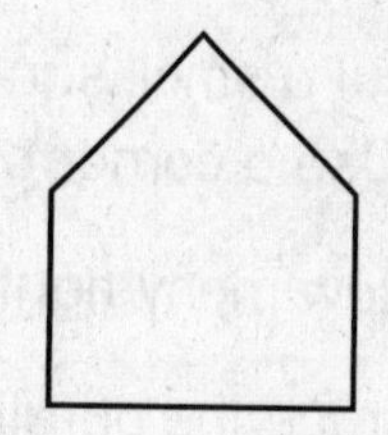

How many right angles are there? ________

Trace the parallel line segments using a red crayon.

5. What numbers would you use to estimate the sum of 23 and 65? ________ and ________

6. Find the answers.

$2 + 3 + 7 + 9 + 1 =$ ________

$$\begin{array}{r} 85 \\ +\ 35 \\ \hline \end{array} \qquad \begin{array}{r} 94 \\ -\ 25 \\ \hline \end{array} \qquad \begin{array}{r} 327 \\ +\ 485 \\ \hline \end{array}$$

Name ______________________

Date ______________________

Saxon Math 2 (for use with **Lesson 114**)

Workspace

1. There are 45 tennis balls and 11 kickballs for the children to use during recess. The children are using 18 tennis balls. How many tennis balls are not being used?

Number sentence ______________________

Answer ______________________

2. This is a tally to show how many children chose each color. Color the graph to show the colors the children chose.

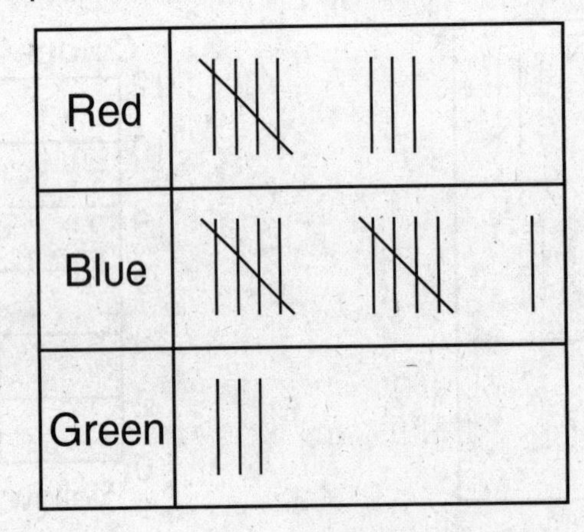

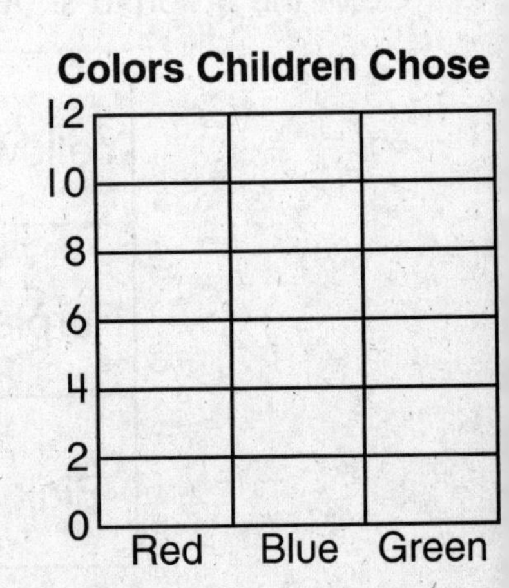

3. Color $3\frac{3}{4}$ squares.

 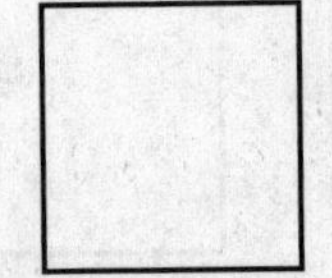 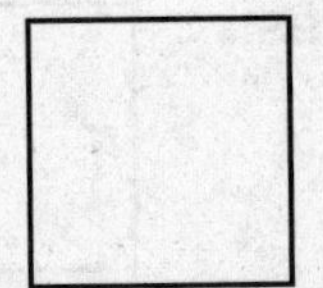

4. Put a small square in each right angle of this shape. (Use a corner of a piece of paper to check the angles.)

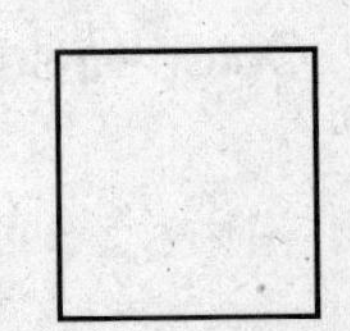

How many right angles are there? ______

Trace the parallel line segments using a red crayon.

5. What numbers would you use to estimate the sum of 17 and 75? ______ and ______

6. Find the answers.

$8 + 3 + 2 + 9 + 1 =$ ______

$$\begin{array}{r} 93 \\ +\ 47 \\ \hline \end{array} \qquad \begin{array}{r} 82 \\ -\ 57 \\ \hline \end{array} \qquad \begin{array}{r} 277 \\ +\ 526 \\ \hline \end{array}$$

Name ______________________ Score ________

Fact Assessment 22-1

Saxon *Math 2 (for use with* ***Lesson 115-1****)*

Set 22: Multiplying by 5

5×3	5×5	5×0	5×7	5×2
5×9	5×1	5×6	5×4	5×8
5×2	5×7	5×4	5×9	5×0
5×5	5×1	5×8	5×3	5×6
5×4	5×0	5×2	5×6	5×5

Name ______________________________

***Saxon** Math 2 (for use with **Lesson 115-1**)*

Set 22: Multiplying by 5

Pretend you are the teacher.
Correct this paper.
If the answer is incorrect, write the correct answer next to the problem.

$\begin{array}{r} 5 \\ \times\ 3 \\ \hline 15 \end{array}$	$\begin{array}{r} 5 \\ \times\ 5 \\ \hline 25 \end{array}$	$\begin{array}{r} 5 \\ \times\ 0 \\ \hline 0 \end{array}$	$\begin{array}{r} 5 \\ \times\ 7 \\ \hline 30 \end{array}$	$\begin{array}{r} 5 \\ \times\ 2 \\ \hline 10 \end{array}$
$\begin{array}{r} 5 \\ \times\ 9 \\ \hline 45 \end{array}$	$\begin{array}{r} 5 \\ \times\ 1 \\ \hline 5 \end{array}$	$\begin{array}{r} 5 \\ \times\ 6 \\ \hline 30 \end{array}$	$\begin{array}{r} 5 \\ \times\ 4 \\ \hline 20 \end{array}$	$\begin{array}{r} 5 \\ \times\ 8 \\ \hline 45 \end{array}$
$\begin{array}{r} 5 \\ \times\ 2 \\ \hline 10 \end{array}$	$\begin{array}{r} 5 \\ \times\ 7 \\ \hline 35 \end{array}$	$\begin{array}{r} 5 \\ \times\ 4 \\ \hline 20 \end{array}$	$\begin{array}{r} 5 \\ \times\ 9 \\ \hline 45 \end{array}$	$\begin{array}{r} 5 \\ \times\ 0 \\ \hline 0 \end{array}$
$\begin{array}{r} 5 \\ \times\ 5 \\ \hline 25 \end{array}$	$\begin{array}{r} 5 \\ \times\ 1 \\ \hline 5 \end{array}$	$\begin{array}{r} 5 \\ \times\ 8 \\ \hline 40 \end{array}$	$\begin{array}{r} 5 \\ \times\ 3 \\ \hline 20 \end{array}$	$\begin{array}{r} 5 \\ \times\ 6 \\ \hline 30 \end{array}$
$\begin{array}{r} 5 \\ \times\ 4 \\ \hline 20 \end{array}$	$\begin{array}{r} 5 \\ \times\ 0 \\ \hline 5 \end{array}$	$\begin{array}{r} 5 \\ \times\ 2 \\ \hline 10 \end{array}$	$\begin{array}{r} 5 \\ \times\ 6 \\ \hline 30 \end{array}$	$\begin{array}{r} 5 \\ \times\ 5 \\ \hline 35 \end{array}$

Name ______________________________

Saxon Math 2 (for use with **Lesson 115-1**)

S100: 100 Subtraction Facts   Corrected by ________

1	$7 - 1$	$10 - 4$	$9 - 0$	$16 - 9$	$5 - 4$	$12 - 6$	$9 - 7$	$11 - 3$	$8 - 2$	$6 - 6$
2	$9 - 4$	$6 - 3$	$11 - 6$	$10 - 2$	$6 - 1$	$12 - 8$	$2 - 0$	$9 - 3$	$7 - 2$	$6 - 5$
3	$5 - 5$	$11 - 4$	$4 - 2$	$15 - 9$	$8 - 0$	$10 - 6$	$14 - 5$	$9 - 9$	$4 - 0$	$12 - 7$
4	$9 - 5$	$17 - 9$	$8 - 4$	$13 - 8$	$9 - 2$	$11 - 5$	$15 - 6$	$5 - 1$	$8 - 5$	$16 - 8$
5	$8 - 6$	$11 - 7$	$1 - 0$	$7 - 3$	$9 - 6$	$4 - 3$	$17 - 8$	$10 - 5$	$12 - 4$	$13 - 7$
6	$8 - 3$	$16 - 7$	$10 - 3$	$4 - 1$	$6 - 2$	$13 - 5$	$7 - 0$	$14 - 9$	$11 - 2$	$10 - 8$
7	$13 - 9$	$10 - 7$	$18 - 9$	$14 - 6$	$1 - 1$	$12 - 3$	$7 - 5$	$2 - 1$	$11 - 8$	$7 - 7$
8	$2 - 2$	$12 - 5$	$3 - 1$	$15 - 7$	$10 - 1$	$6 - 0$	$13 - 4$	$5 - 2$	$9 - 8$	$3 - 0$
9	$11 - 9$	$7 - 6$	$13 - 6$	$3 - 3$	$14 - 8$	$9 - 1$	$6 - 4$	$12 - 9$	$7 - 4$	$8 - 7$
10	$4 - 4$	$15 - 8$	$3 - 2$	$5 - 0$	$5 - 3$	$8 - 8$	$14 - 7$	$10 - 9$	$0 - 0$	$8 - 1$

Name ____________________

Lesson Worksheet 115-1A

***Saxon** Math 2 (for use with **Lesson 115-1**)*

1. Fill in the products.

$0 \times 2 =$ $5 \times 2 =$

$1 \times 2 =$ $6 \times 2 =$

$2 \times 2 =$ $7 \times 2 =$

$3 \times 2 =$ $8 \times 2 =$

$4 \times 2 =$ $9 \times 2 =$

3. Fill in the missing factors.

$\square \times 2 = 8$

$\square \times 2 = 18$

$\square \times 2 = 2$

$\square \times 2 = 14$

$\square \times 2 = 6$

2. Match the problem to the answers.

5×2 •	• 2
1×2 •	• 6
8×2 •	• 10
3×2 •	• 0
6×2 •	• 16
0×2 •	• 18
9×2 •	• 12
2×2 •	• 8
7×2 •	• 14
4×2 •	• 4

Name ____________________

Saxon Math 2 (for use with **Lesson 115-1**)

Name

Draw an 8-cm line segment.

Date

Measure this line segment using centimeters. ________ cm

1. The children were walking in pairs. George counted eight pairs of children. Draw x's to show the children.

How many children is that? ____________

2. Use the graph to answer the questions.

How many children chose skating? ________

Color the graph to show that 4 children chose skiing.

How many more children chose biking than skiing? ________

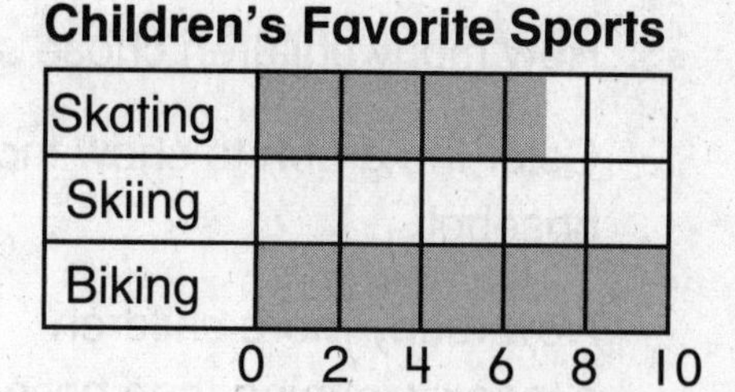

3. Show 7:42 on the clock.

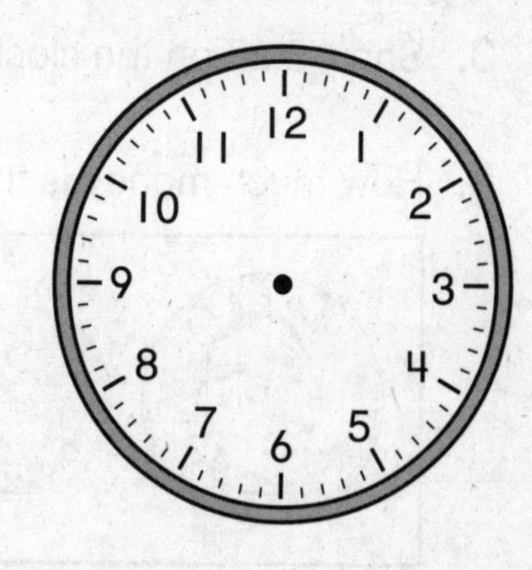

4. How much money is this? ____________

5. Complete the number patterns.

______, ______, ______, ______, ______, 703, 704, 705

______, ______, ______, 425, 435, 445, ______, ______, ______

6. Find the answers.

$7 \times 2 =$ ______ $8 \times 5 =$ ______

$9 \times 2 =$ ______ $5 \times 2 =$ ______

$$\begin{array}{r} 80 \\ -\ 23 \\ \hline \end{array} \qquad \begin{array}{r} 742 \\ +\ 138 \\ \hline \end{array}$$

Name ______________________________

Date ______________________________

***Saxon** Math 2 (for use with **Lesson 115-1**)*

1. The children were playing in pairs. Linda counted four pairs of children. Draw x's to show the children.

How many children is that? ____________

2. Use the graph to answer the questions.

How many children chose soccer? ________

Color the graph to show that 2 children chose baseball.

How many more children chose swimming than baseball? ________

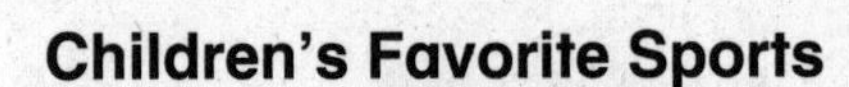

3. Show 3:27 on the clock.

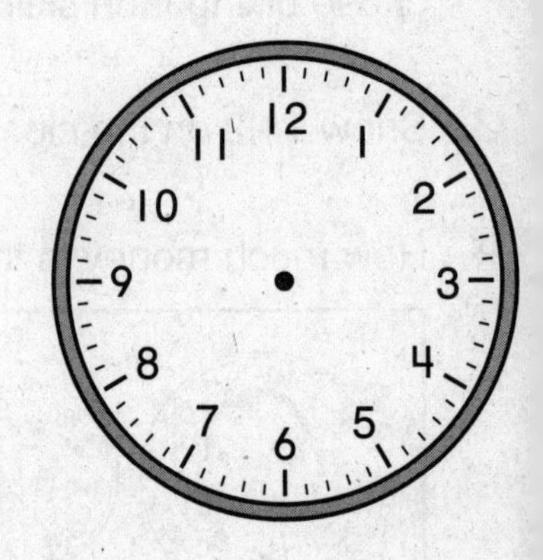

4. How much money is this? ____________

5. Complete the number patterns.

______, ______, ______, ______, ______, 402, 403, 404

______, ______, ______, 647, 637, 627, ______, ______, ______

6. Find the answers.

$9 \times 2 =$ ______ $\quad$ $6 \times 5 =$ ______

$8 \times 2 =$ ______ $\quad$ $7 \times 5 =$ ______

$$\begin{array}{r} 90 \\ -\ 48 \\ \hline \end{array} \qquad \begin{array}{r} 275 \\ +\ 516 \\ \hline \end{array}$$

Name ______________________ Score ______

Fact Assessment 22-2

***Saxon** Math 2 (for use with **Lesson 115-2**)*

S100: 100 Subtraction Facts

1	$7 - 1$	$10 - 4$	$9 - 0$	$16 - 9$	$5 - 4$	$12 - 6$	$9 - 7$	$11 - 3$	$8 - 2$	$6 - 6$
2	$9 - 4$	$6 - 3$	$11 - 6$	$10 - 2$	$6 - 1$	$12 - 8$	$2 - 0$	$9 - 3$	$7 - 2$	$6 - 5$
3	$5 - 5$	$11 - 4$	$4 - 2$	$15 - 9$	$8 - 0$	$10 - 6$	$14 - 5$	$9 - 9$	$4 - 0$	$12 - 7$
4	$9 - 5$	$17 - 9$	$8 - 4$	$13 - 8$	$9 - 2$	$11 - 5$	$15 - 6$	$5 - 1$	$8 - 5$	$16 - 8$
5	$8 - 6$	$11 - 7$	$1 - 0$	$7 - 3$	$9 - 6$	$4 - 3$	$17 - 8$	$10 - 5$	$12 - 4$	$13 - 7$
6	$8 - 3$	$16 - 7$	$10 - 3$	$4 - 1$	$6 - 2$	$13 - 5$	$7 - 0$	$14 - 9$	$11 - 2$	$10 - 8$
7	$13 - 9$	$10 - 7$	$18 - 9$	$14 - 6$	$1 - 1$	$12 - 3$	$7 - 5$	$2 - 1$	$11 - 8$	$7 - 7$
8	$2 - 2$	$12 - 5$	$3 - 1$	$15 - 7$	$10 - 1$	$6 - 0$	$13 - 4$	$5 - 2$	$9 - 8$	$3 - 0$
9	$11 - 9$	$7 - 6$	$13 - 6$	$3 - 3$	$14 - 8$	$9 - 1$	$6 - 4$	$12 - 9$	$7 - 4$	$8 - 7$
10	$4 - 4$	$15 - 8$	$3 - 2$	$5 - 0$	$5 - 3$	$8 - 8$	$14 - 7$	$10 - 9$	$0 - 0$	$8 - 1$

Name ______________________________

Date ______________________________

1. Alex had 36 nickels and 83 pennies. He gave 27 pennies to Marcus. How many pennies does he have now?

Workspace

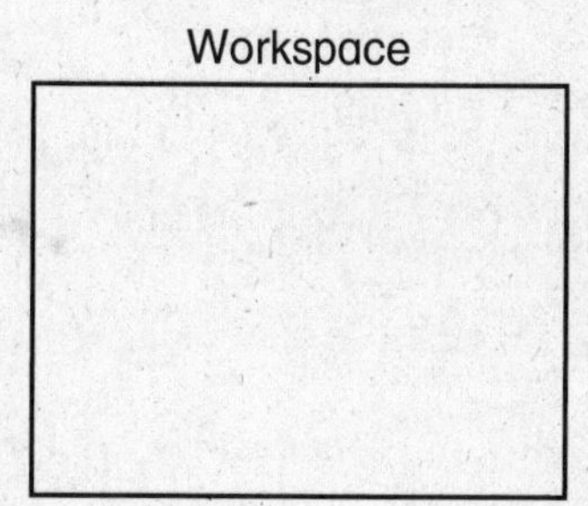

Number sentence ______________________________

Answer ______________________________

2. Show how to share the balloons equally.

12 balloons

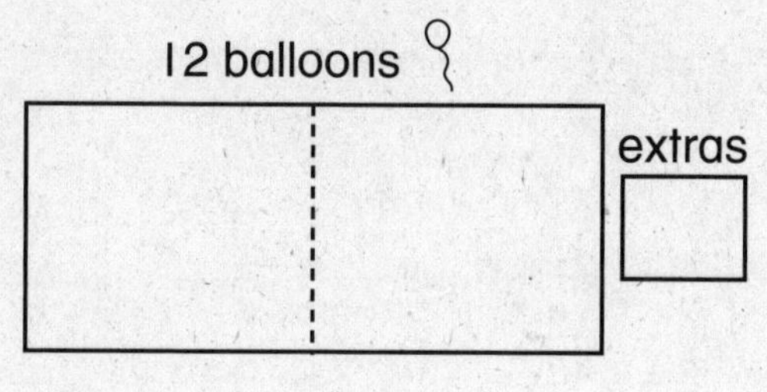

extras

9 balloons

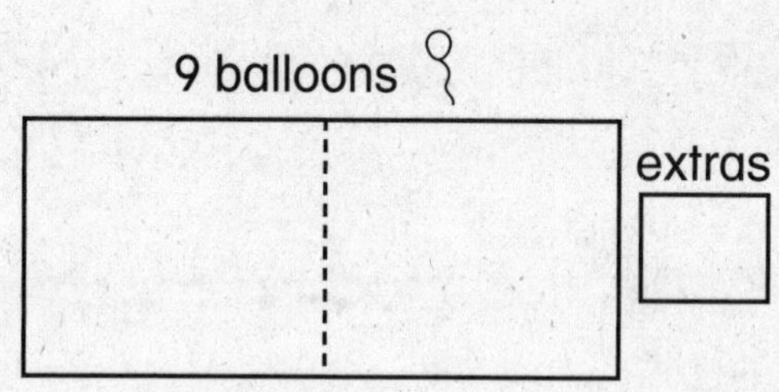

extras

One half of 12 is ________

One half of 9 is ________

3. How much money is this?
Write the amount two ways.

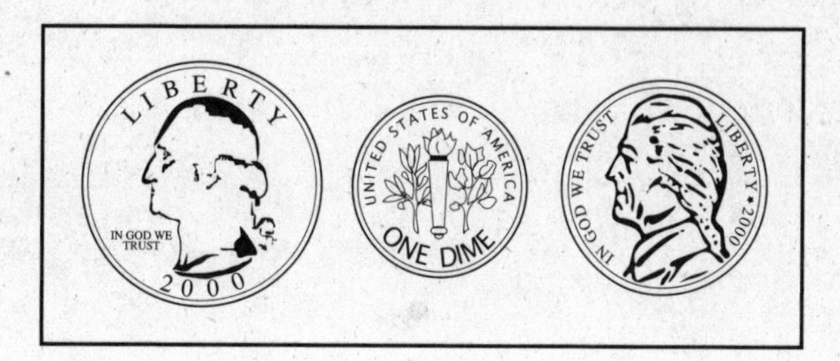

__________ __________

4. Use a crayon to trace an example of parallel lines on this paper.

Where do you see parallel lines in the classroom?

____

5. Show eleven twenty-four on both clocks.

What time is shown on this clock?

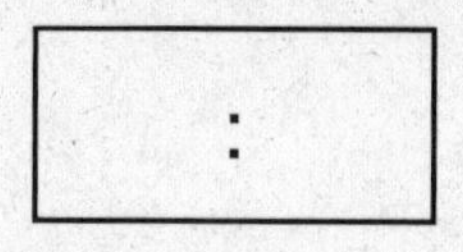

6. Find the answers.

$$\begin{array}{r} 79 \\ +\ 53 \\ \hline \end{array} \qquad \begin{array}{r} 61 \\ -\ 25 \\ \hline \end{array}$$

86 + 54

$$\begin{array}{r} \\ + \quad \\ \hline \end{array}$$

84 – 57

$$\begin{array}{r} \\ - \quad \\ \hline \end{array}$$

Name ________________________________

Lesson Worksheet 115-2

Saxon *Math 2 (for use with* ***Lesson 115-2****)*

Use 1" square color tiles to cover each rectangle.

A.

B.

C.

D.

Area of each rectangle

A. ________ 1" square color tiles

B. ________ 1" square color tiles

C. ________ 1" square color tiles

D. ________ 1" square color tiles

Name ______________________________

Class Fact Practice 116A

***Saxon** Math 2 (for use with **Lesson 116**)*

Set 23: Multiplying by 2

2×0	2×1	2×2	2×3	2×4
2×5	2×6	2×7	2×8	2×9
2×6	2×2	2×8	2×0	2×4
2×5	2×1	2×9	2×7	2×3
2×4	2×6	2×5	2×8	2×9

Name ____________________

Set 23: Multiplying by 2

1. Read the answers to someone.
2. Write the answers.
3. Ask someone to correct your paper. Corrected by ____________________

2×0	2×1	2×2	2×3	2×4
2×5	2×6	2×7	2×8	2×9
2×4	2×1	2×8	2×2	2×5
2×3	2×9	2×0	2×6	2×7
2×8	2×4	2×6	2×3	2×9

Name ____________________

Lesson Worksheet 116A

***Saxon** Math 2 (for use with **Lesson 116**)*

1.

___ groups of ___ wheels is ___ wheels

___ × ___ wheels = ___ wheels

2.

___ groups of ___ buttons is ___ buttons

___ × ___ buttons = ___ buttons

3.

___ groups of ___ hearts is ___ hearts

___ × ___ hearts = ___ hearts

4.

___ groups of ___ wheels is ___ wheels

___ × ___ wheels = ___ wheels

Name ______________________

5.

___ groups of ___ crackers is ___ crackers

___ × ___ crackers = ___ crackers

6.

___ × ______________ = ______________

7.

___ × ______________ = ______________

8.

___ × ______________ = ______________

Name •

Draw an 8-cm line segment.

Date •————————•

Measure this line segment using centimeters. ________ cm

1. Kyle tallied the number of children who wore green.

How many children wore green? ________

Twice as many children wore green as yellow.

How many children wore yellow? ________

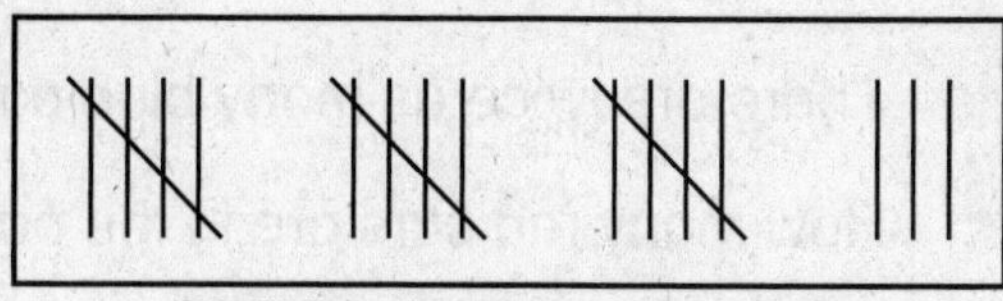

Children Wearing Green

2. Show 4:53 on the clock.

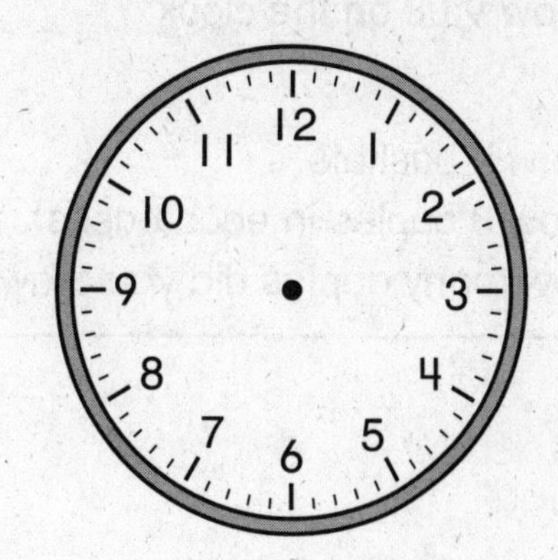

3. Draw 3 baskets.
Draw 4 oranges in each basket.
How many oranges did you draw? ________

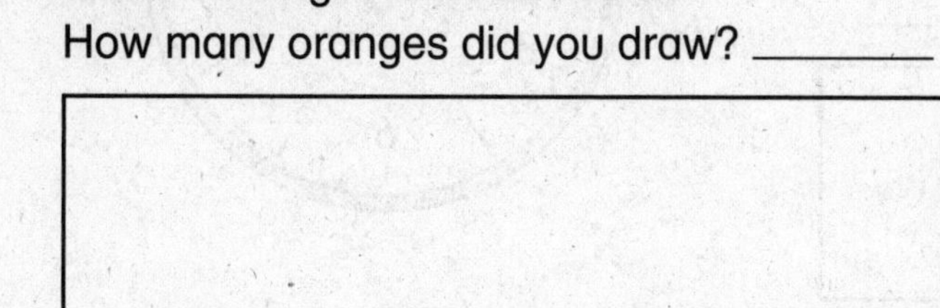

4. Draw a small square to show the right angle in the triangle.

5. I have 2 quarters, 1 dime, 3 nickels, and 4 pennies. Draw the coins.

How much money do I have? ________________

6. Use the correct symbol (+, –, or ×).

4 ◯ 2 = 8 2 ◯ 5 = 7

7. Find the answers.

$$\begin{array}{r} 86 \\ -\ 72 \\ \hline \end{array} \qquad \begin{array}{r} 41 \\ -\ 17 \\ \hline \end{array} \qquad \begin{array}{r} 621 \\ +\ 189 \\ \hline \end{array} \qquad \begin{array}{r} \$3.79 \\ +\ 2.43 \\ \hline \end{array}$$

Name ______________________________

Date ______________________________

1. Moriah tallied the number of blue cars in the parking lot.

How many blue cars did Moriah count? ________

There are twice as many blue cars as red cars.

How many red cars are in the parking lot? ________

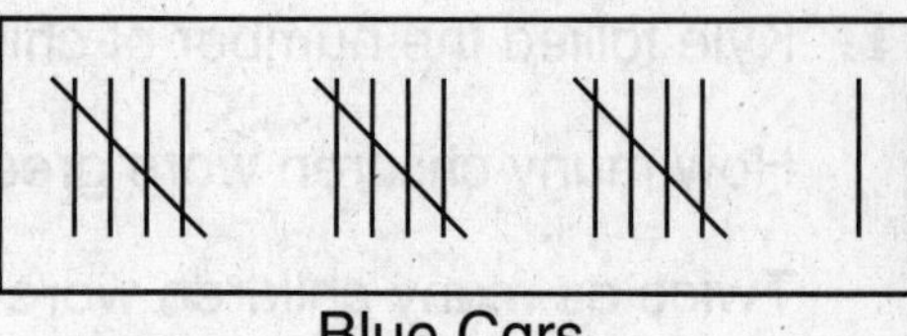

Blue Cars

2. Show 9:08 on the clock.

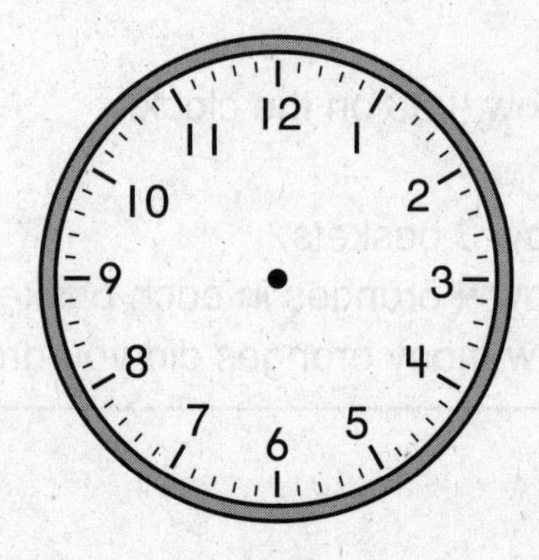

3. Draw 4 baskets.
Draw 3 apples in each basket.
How many apples did you draw? ________

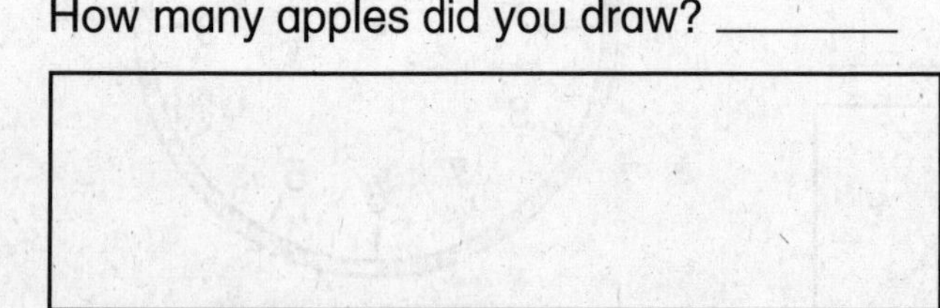

4. Draw a small square to show the right angle in the triangle.

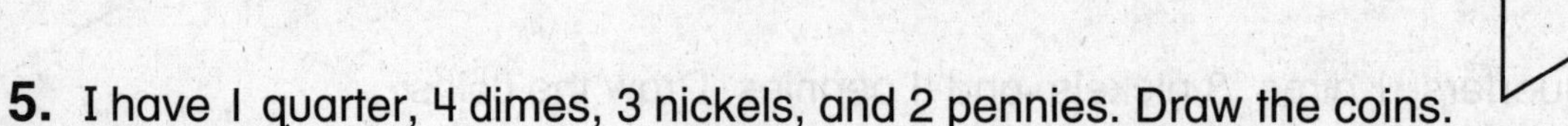

5. I have 1 quarter, 4 dimes, 3 nickels, and 2 pennies. Draw the coins.

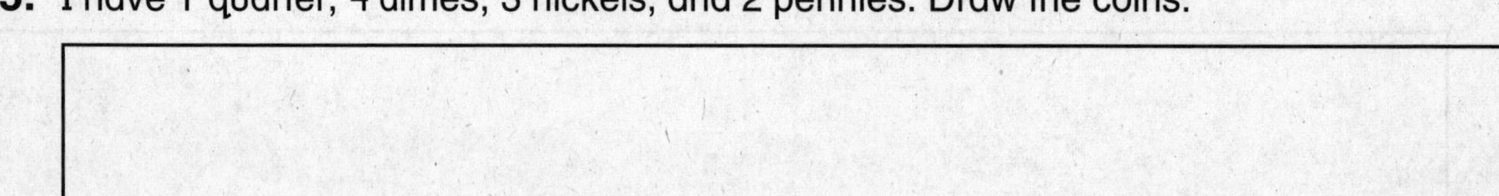

How much money do I have? ____________________

6. Use the correct symbol (+, −, or ×).

$4 \bigcirc 5 = 20$ $\qquad$ $5 \bigcirc 2 = 10$

7. Find the answers.

$$\begin{array}{r} 78 \\ -\ 37 \\ \hline \end{array} \qquad \begin{array}{r} 65 \\ -\ 48 \\ \hline \end{array} \qquad \begin{array}{r} 367 \\ +\ 243 \\ \hline \end{array} \qquad \begin{array}{r} \$6.34 \\ +\ \ 3.18 \\ \hline \end{array}$$

Name ____________________

*Saxon Math 2 (for use with **Lesson 117**)*

Set 23: Multiplying by 2   Corrected by ______

1. Fill in the products.

$0 \times 2 =$ $5 \times 2 =$

$1 \times 2 =$ $6 \times 2 =$

$2 \times 2 =$ $7 \times 2 =$

$3 \times 2 =$ $8 \times 2 =$

$4 \times 2 =$ $9 \times 2 =$

2. Match the problems to the answers.

3×2 •	• 14
7×2 •	• 6
5×2 •	• 16
2×2 •	• 2
8×2 •	• 10
1×2 •	• 4
6×2 •	• 0
4×2 •	• 18
0×2 •	• 12
9×2 •	• 8

3. Fill in the missing factors.

$\square \times 2 = 12$

$\square \times 2 = 4$

$\square \times 2 = 16$

$\square \times 2 = 0$

$\square \times 2 = 10$

Name .
Draw a 7-cm line segment.

Date •———————————•
Measure this line segment using centimeters. ________ cm

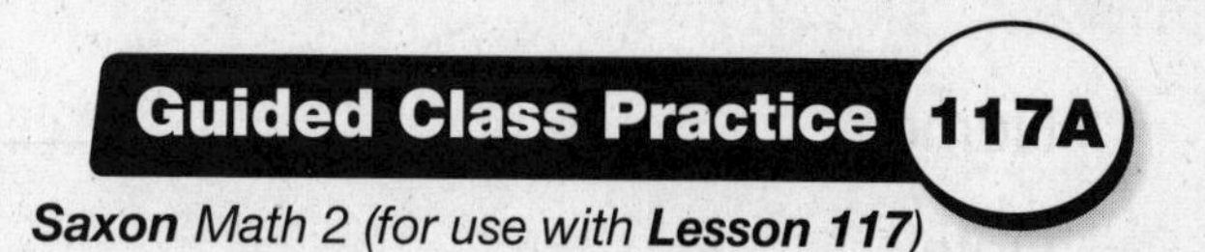

Saxon Math 2 (for use with **Lesson 117**)

1. There were 6 children at the party. Mrs. Parsons put 5 strawberries on each child's dish of ice cream. Draw a picture to show the strawberries on the dishes of ice cream.

What type of story problem is this? ____________

How many strawberries did Mrs. Parsons use altogether?

Number sentence ____________

Answer ____________

2. Use the graph to answer the questions.

How many children chose winter? ________

How many more children chose spring than fall? ________

Write one fact about the information on the graph.

____

____

Children's Favorite Seasons

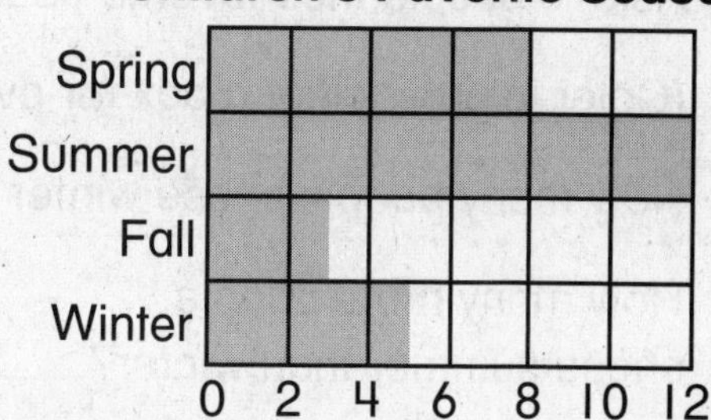

3. Circle the letters that have parallel line segments.

A	E	L	M	V	Z

4. Write a mixed number to show how much is shaded.

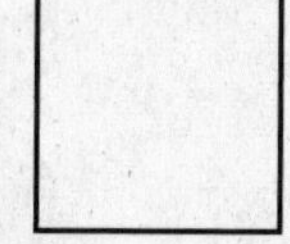

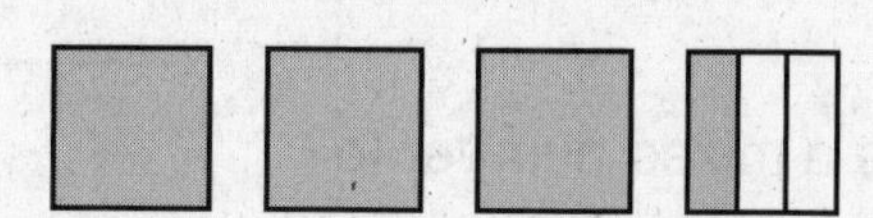

5. Write the products.

2	2	2	2	2	2	2	2	2	2	2
× 4	× 8	× 3	× 7	× 5	× 9	× 1	× 6	× 2	× 0	× 10

Name ______________________

Date ______________________

Saxon Math 2 (for use with ***Lesson 117****)*

1. There were 8 children in the Reading Room. Mrs. Kennedy put 2 stickers on each child's paper. Draw a picture to show the stickers on the papers.

What type of story problem is this? ______________

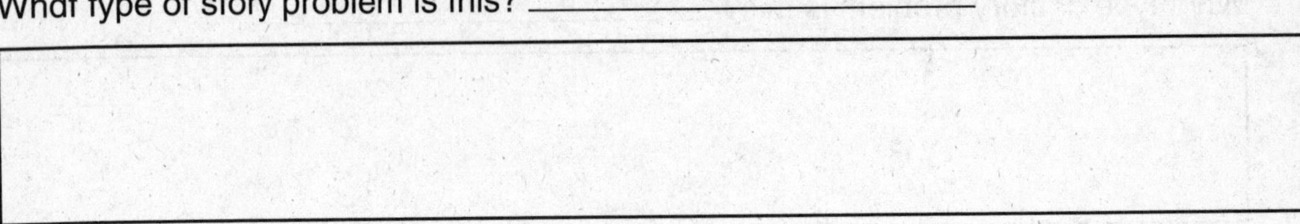

How many stickers did Mrs. Kennedy use altogether?

Number sentence ______________________

Answer ______________

2. Ask 10 people their favorite season.

(Color in one half of a box for every vote.)

How many people chose winter? ________

How many more people chose summer than winter? ________

Write one fact about the information on the graph.

____

____

3. Circle the letters that have parallel line segments.

H Y I N T X

4. Write a mixed number to show how much is shaded.

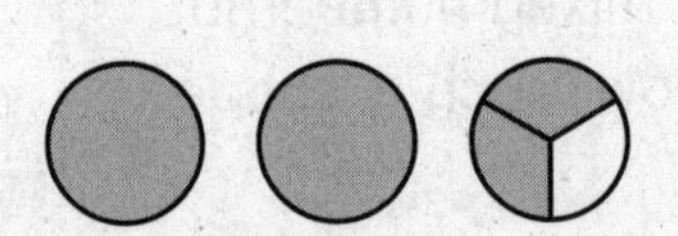

5. Fill in the missing factors.

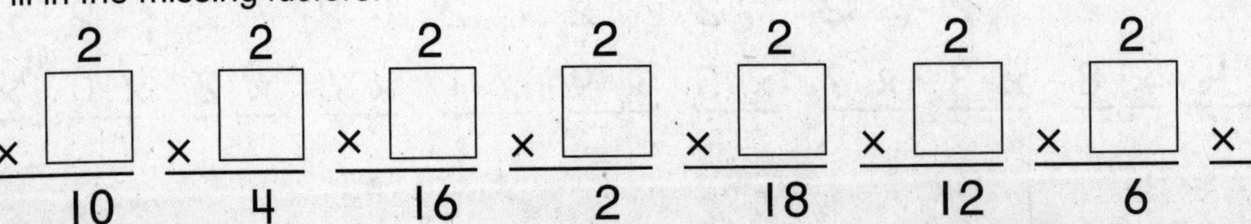

Name ______________________________

Perpendicular Line Segments

1.

2.

3.

4.

Name •
Draw a 7-cm line segment.

Guided Class Practice 118A

***Saxon** Math 2 (for use with **Lesson 118**)*

Date •————————•
Measure this line segment using centimeters. ________ cm

1. Pencils are sold in packages of 3. Mrs. Conlan bought 7 packages of pencils. Draw a picture to show the packages of pencils.

What type of story problem is this? ____________________

How many pencils did she buy?

Number sentence ________________________________

Answer ____________________

2. Circle the perpendicular line segments.

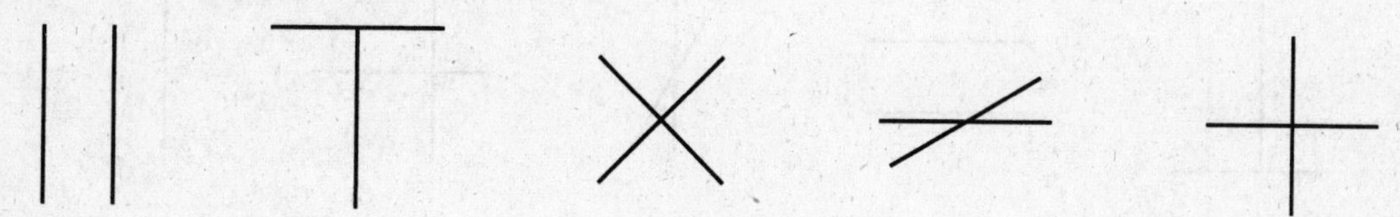

3. How many small squares are in this rectangle?

Area = ________ square units

Color one square.

What fractional part of the rectangle is colored? ____

4. Round each number to the nearest 10.

78 ______ 13 ______ 25 ______

5. Circle all the geometric solids that have at least one point (vertex).

pyramid cylinder cone sphere cube

6. Find the answers.

```
  62        68       $2.93
- 38        37     +  3.78
----      + 25     -------
          ----
```

$8 \times 10 =$ ______

$3 \times 100 =$ ______

Name ______________________________

Date ______________________________

Homework 118B

***Saxon** Math 2 (for use with **Lesson 118**)*

1. Markers are sold in packages of 10. Mrs. Campion bought 4 packages. Draw a picture to show the packages of markers.

What type of story problem is this? ____________________

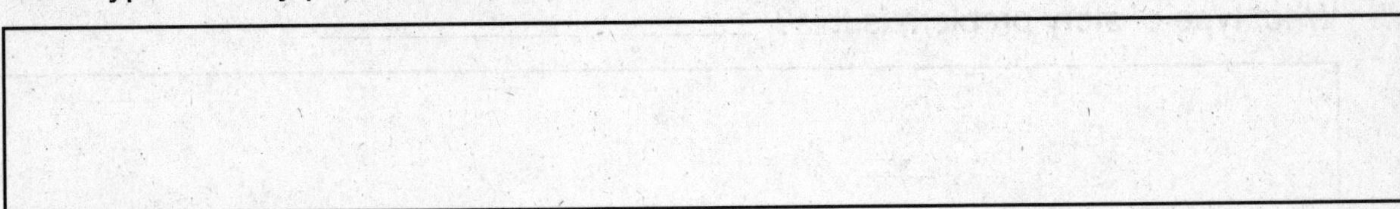

How many markers did she buy?

Number sentence ______________________________

Answer ____________________

2. Circle the perpendicular line segments.

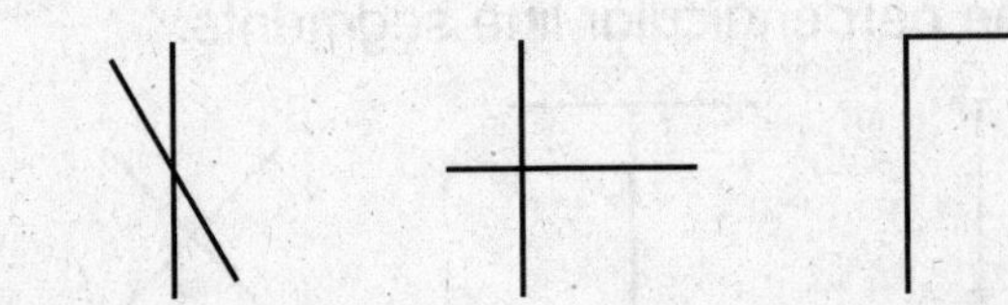

3. How many small squares are in this large square?

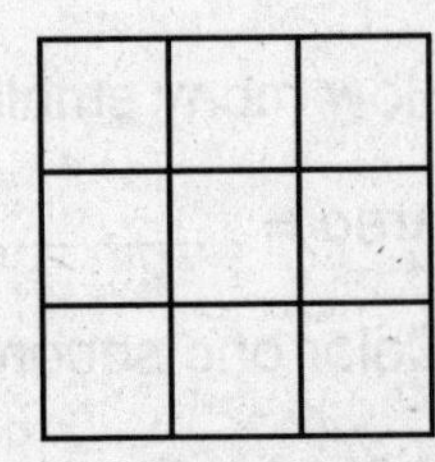

Area = ________ square units

Color one square.

What fractional part of the large square is colored? ______

4. Round each number to the nearest 10.

31 ________ 9 ________ 15 ________

5. Circle all the geometric solids that will roll.

pyramid cylinder cone sphere

6. Find the answers.

$$\begin{array}{r} 71 \\ -\ 25 \\ \hline \end{array} \qquad \begin{array}{r} 79 \\ 26 \\ +\ 35 \\ \hline \end{array} \qquad \begin{array}{r} \$2.56 \\ +\ 4.74 \\ \hline \end{array}$$

$7 \times 10 =$ ________

$2 \times 100 =$ ________

Name ____________________

Class Fact Practice 119A

Saxon *Math 2 (for use with* ***Lesson 119****)*

Set 23: Multiplying by 2

2×3	2×9	2×4	2×1	2×6
2×8	2×0	2×5	2×7	2×2
2×6	2×4	2×9	2×1	2×8
2×3	2×7	2×2	2×5	2×0
2×7	2×1	2×4	2×8	2×3

Name ______________________

Saxon Math 2 (for use with **Lesson 119**)

Set 23: Multiplying by 2

1. Read the answers to someone.
2. Write the answers.
3. Ask someone to correct your paper. Corrected by ______________

2×7	2×0	2×5	2×9	2×1
2×4	2×8	2×2	2×6	2×3
2×6	2×4	2×1	2×8	2×2
2×0	2×7	2×5	2×3	2×9
2×4	2×8	2×2	2×7	2×3

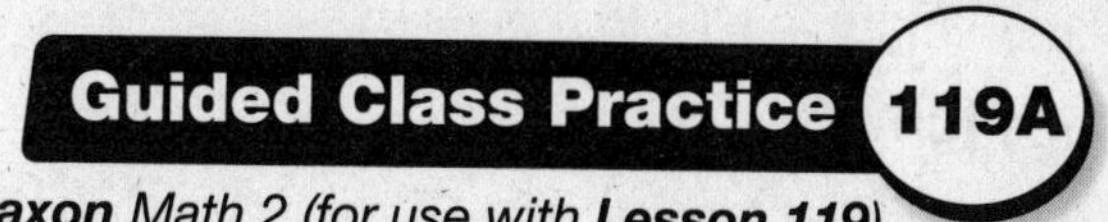

Guided Class Practice 119A

Saxon Math 2 (for use with **Lesson 119**)

Name

Draw an 8-cm line segment.

Date

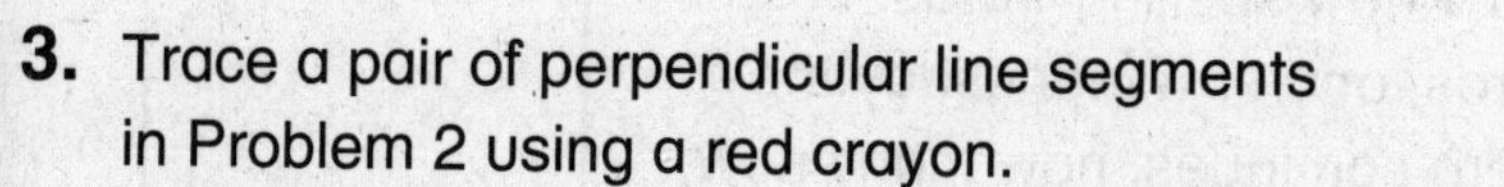

Measure this line segment using centimeters. ________ cm

1. There are five desks in the room. Ashleigh put three books on each desk. Draw the books on the desks. How many books are there altogether?

Number sentence ______________________________

Answer ______________________

2. Measure the vertical line segment on the left using centimeters. ________ cm

Measure the vertical line segment on the right using centimeters. ________ cm

Measure the horizontal line segment using centimeters. ________ cm

Measure the oblique line segment using centimeters. ________ cm

What is the perimeter of the shape? ________ cm

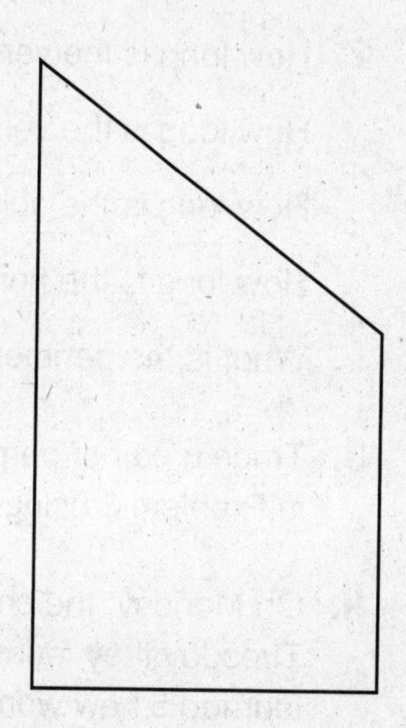

3. Trace a pair of perpendicular line segments in Problem 2 using a red crayon.

4. On Monday, the children read 2 pages, on Tuesday they read 4 pages, and on Wednesday they read 6 pages. If the pattern continues, how many pages will they read on Friday? ____________

Monday	Tuesday	Wednesday	Thursday	Friday
2 pages	4 pages	6 pages		

5. Write six hundred seventeen using digits. ________

Write this number in expanded form. ______________________________

6. Find the answers.

$$\begin{array}{r} 362 \\ -\ 125 \\ \hline \end{array} \qquad \begin{array}{r} 549 \\ -\ 365 \\ \hline \end{array} \qquad \begin{array}{r} \$6.43 \\ -\ \ 4.26 \\ \hline \end{array} \qquad \begin{array}{r} \$3.12 \\ +\ \ 6.29 \\ \hline \end{array}$$

Name ______________________

Date ______________________

Saxon *Math 2 (for use with* ***Lesson 119****)*

1. There are two tables in the room. Brenden put six books on each table. Draw the books on the tables. How many books are there altogether?

Number sentence ______________________

Answer ______________________

2. How long is the vertical line segment on the right? ______ cm

How long is the vertical line segment on the left? ______ cm

How long is the oblique line segment? ______ cm

How long is the horizontal line segment? ______ cm

What is the perimeter of the shape? ______ cm

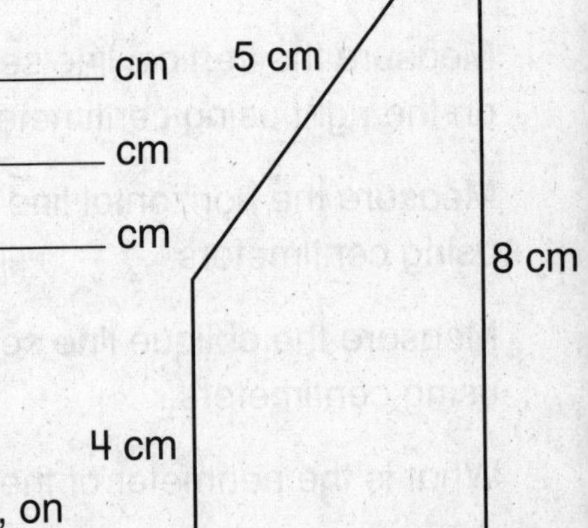

3. Trace a pair of perpendicular line segments in Problem 2 using a red crayon.

4. On Monday, the children learned 9 new spelling words, on Tuesday they learned 7 new words, and on Wednesday they learned 5 new words. If the pattern continues, how many new words will they learn on Friday? ______

Monday	Tuesday	Wednesday	Thursday	Friday
9 words	7 words	5 words		

5. Write two hundred thirty-seven using digits. ______

Write this number in expanded form. ______

6. Find the answers.

$$\begin{array}{r} 592 \\ -\ 314 \\ \hline \end{array} \qquad \begin{array}{r} 687 \\ -\ 294 \\ \hline \end{array} \qquad \begin{array}{r} \$5.42 \\ -\ 2.70 \\ \hline \end{array} \qquad \begin{array}{r} \$7.24 \\ +\ 2.38 \\ \hline \end{array}$$

Name ______________________ Score ________

Fact Assessment 23-1

***Saxon** Math 2 (for use with **Lesson 120-1**)*

Set 23: Multiplying by 2

2×3	2×5	2×0	2×7	2×2
2×9	2×1	2×6	2×4	2×8
2×2	2×7	2×4	2×9	2×0
2×5	2×1	2×8	2×3	2×6
2×4	2×0	2×2	2×6	2×5

Name ____________________

Fact Homework 120A

Saxon *Math 2 (for use with* ***Lesson 120-1****)*

Set 23: Multiplying by 2

Pretend you are the teacher.
Correct this paper.
If the answer is incorrect, write the correct answer next to the problem.

$\begin{array}{r} 2 \\ \times\ 3 \\ \hline 6 \end{array}$	$\begin{array}{r} 2 \\ \times\ 5 \\ \hline 10 \end{array}$	$\begin{array}{r} 2 \\ \times\ 0 \\ \hline 2 \end{array}$	$\begin{array}{r} 2 \\ \times\ 7 \\ \hline 14 \end{array}$	$\begin{array}{r} 2 \\ \times\ 2 \\ \hline 4 \end{array}$
$\begin{array}{r} 2 \\ \times\ 9 \\ \hline 18 \end{array}$	$\begin{array}{r} 2 \\ \times\ 1 \\ \hline 2 \end{array}$	$\begin{array}{r} 2 \\ \times\ 6 \\ \hline 14 \end{array}$	$\begin{array}{r} 2 \\ \times\ 4 \\ \hline 8 \end{array}$	$\begin{array}{r} 2 \\ \times\ 8 \\ \hline 16 \end{array}$
$\begin{array}{r} 2 \\ \times\ 2 \\ \hline 4 \end{array}$	$\begin{array}{r} 2 \\ \times\ 7 \\ \hline 14 \end{array}$	$\begin{array}{r} 2 \\ \times\ 4 \\ \hline 8 \end{array}$	$\begin{array}{r} 2 \\ \times\ 9 \\ \hline 18 \end{array}$	$\begin{array}{r} 2 \\ \times\ 0 \\ \hline 0 \end{array}$
$\begin{array}{r} 2 \\ \times\ 5 \\ \hline 10 \end{array}$	$\begin{array}{r} 2 \\ \times\ 1 \\ \hline 2 \end{array}$	$\begin{array}{r} 2 \\ \times\ 8 \\ \hline 14 \end{array}$	$\begin{array}{r} 2 \\ \times\ 3 \\ \hline 6 \end{array}$	$\begin{array}{r} 2 \\ \times\ 6 \\ \hline 12 \end{array}$
$\begin{array}{r} 2 \\ \times\ 4 \\ \hline 8 \end{array}$	$\begin{array}{r} 2 \\ \times\ 0 \\ \hline 0 \end{array}$	$\begin{array}{r} 2 \\ \times\ 2 \\ \hline 4 \end{array}$	$\begin{array}{r} 2 \\ \times\ 6 \\ \hline 12 \end{array}$	$\begin{array}{r} 2 \\ \times\ 5 \\ \hline 10 \end{array}$

Name ____________________

Saxon Math 2 (for use with **Lesson 120-1**)

S100: 100 Subtraction Facts   Corrected by ________

Row										
1	7 − 1	10 − 4	9 − 0	16 − 9	5 − 4	12 − 6	9 − 7	11 − 3	8 − 2	6 − 6
2	9 − 4	6 − 3	11 − 6	10 − 2	6 − 1	12 − 8	2 − 0	9 − 3	7 − 2	6 − 5
3	5 − 5	11 − 4	4 − 2	15 − 9	8 − 0	10 − 6	14 − 5	9 − 9	4 − 0	12 − 7
4	9 − 5	17 − 9	8 − 4	13 − 8	9 − 2	11 − 5	15 − 6	5 − 1	8 − 5	16 − 8
5	8 − 6	11 − 7	1 − 0	7 − 3	9 − 6	4 − 3	17 − 8	10 − 5	12 − 4	13 − 7
6	8 − 3	16 − 7	10 − 3	4 − 1	6 − 2	13 − 5	7 − 0	14 − 9	11 − 2	10 − 8
7	13 − 9	10 − 7	18 − 9	14 − 6	1 − 1	12 − 3	7 − 5	2 − 1	11 − 8	7 − 7
8	2 − 2	12 − 5	3 − 1	15 − 7	10 − 1	6 − 0	13 − 4	5 − 2	9 − 8	3 − 0
9	11 − 9	7 − 6	13 − 6	3 − 3	14 − 8	9 − 1	6 − 4	12 − 9	7 − 4	8 − 7
10	4 − 4	15 − 8	3 − 2	5 − 0	5 − 3	8 − 8	14 − 7	10 − 9	0 − 0	8 − 1

Name ______________________________

Lesson Worksheet 120-1A

Saxon Math 2 (for use with Lesson 120-1)

1. Fill in the products.

$0 \times 3 =$ $5 \times 3 =$

$1 \times 3 =$ $6 \times 3 =$

$2 \times 3 =$ $7 \times 3 =$

$3 \times 3 =$ $8 \times 3 =$

$4 \times 3 =$ $9 \times 3 =$

2. Match the problems to the answers.

3×3 •	• 18
9×3 •	• 0
6×3 •	• 9
0×3 •	• 12
4×3 •	• 27
8×3 •	• 3
1×3 •	• 24
5×3 •	• 21
7×3 •	• 6
2×3 •	• 15

3. Fill in the missing factors.

$\square \times 3 = 6$

$\square \times 3 = 21$

$\square \times 3 = 3$

$\square \times 3 = 27$

$\square \times 3 = 12$

Name ______________________________

Name ______________________

Date ______________________

Problem-Solving Worksheet 120A

Saxon *Math 2 (for use with* ***Lesson 120-1****)*

Understand	Plan	Solve	Check

Make a Table

At the school store Nicholas can buy a pencil for 3¢. Nicholas has 22¢. Show how many pencils Nicholas can buy with his money.

Number of pencils	1	2	3	4					
Cost	3¢	6¢							

How many pencils can Nicholas buy with his money? ______________

Name ____________________

Understand	Plan	Solve	Check

At the school store Marybeth can buy a folder for 25¢. Marybeth has $1.60. Show how many folders Marybeth can buy with her money.

Number of folders	1	2	3	4				
Cost	25¢	50¢						

How many folders can Marybeth buy with her money? ____________

Circle the problem-solving strategies you used to solve this problem.

Act It Out

Draw a Picture

Make an Organized List

Make a Table

Use Logical Reasoning

Look for a Pattern

Guess and Check

Explain how you got your answer: ____________________

Name ____________________ Score ________

Fact Assessment 23-2

***Saxon** Math 2 (for use with **Lesson 120-2**)*

S100: 100 Subtraction Facts

1	7 − 1	10 − 4	9 − 0	16 − 9	5 − 4	12 − 6	9 − 7	11 − 3	8 − 2	6 − 6
2	9 − 4	6 − 3	11 − 6	10 − 2	6 − 1	12 − 8	2 − 0	9 − 3	7 − 2	6 − 5
3	5 − 5	11 − 4	4 − 2	15 − 9	8 − 0	10 − 6	14 − 5	9 − 9	4 − 0	12 − 7
4	9 − 5	17 − 9	8 − 4	13 − 8	9 − 2	11 − 5	15 − 6	5 − 1	8 − 5	16 − 8
5	8 − 6	11 − 7	1 − 0	7 − 3	9 − 6	4 − 3	17 − 8	10 − 5	12 − 4	13 − 7
6	8 − 3	16 − 7	10 − 3	4 − 1	6 − 2	13 − 5	7 − 0	14 − 9	11 − 2	10 − 8
7	13 − 9	10 − 7	18 − 9	14 − 6	1 − 1	12 − 3	7 − 5	2 − 1	11 − 8	7 − 7
8	2 − 2	12 − 5	3 − 1	15 − 7	10 − 1	6 − 0	13 − 4	5 − 2	9 − 8	3 − 0
9	11 − 9	7 − 6	13 − 6	3 − 3	14 − 8	9 − 1	6 − 4	12 − 9	7 − 4	8 − 7
10	4 − 4	15 − 8	3 − 2	5 − 0	5 − 3	8 − 8	14 − 7	10 − 9	0 − 0	8 − 1

Name ______________________

Date ______________________

1. Mrs. Sarno has 4 packages of markers. Each package has 5 markers. Draw a picture to show the markers. How many markers does Mrs. Sarno have altogether?

Number sentence ______________________

Answer ______________________

2. Measure the length of each side of this shape using centimeters.

What is the perimeter?

______

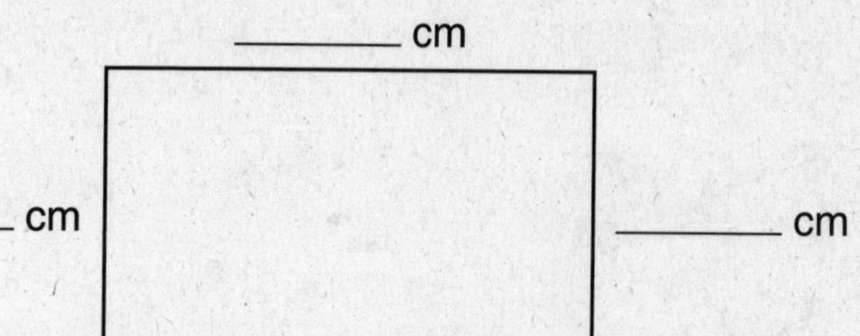

______ cm

______ cm ______ cm

______ cm

3. Draw an 8-cm line segment.

•

4. I have 1 quarter, 2 dimes, and 1 nickel. Draw the coins. How much money do I have? Write the amount two ways.

______ ______

5. Five children have red lunch boxes. Twelve children have yellow lunch boxes. Nine children have blue lunch boxes.

Shade the graph to show the number of children with each color lunch box.

Lunch Box Colors

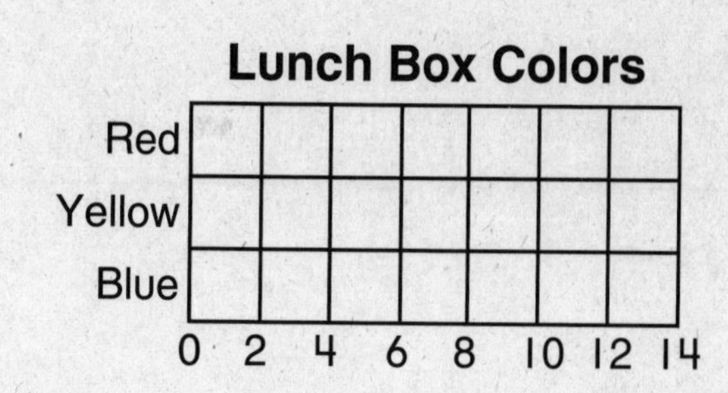

6. Find the answers.

$3 \times 5 =$ ______

$$\begin{array}{r} 342 \\ +\,486 \\ \hline \end{array} \quad \begin{array}{r} 45 \\ -\,29 \\ \hline \end{array} \quad \begin{array}{r} 70 \\ -\,34 \\ \hline \end{array} \quad \begin{array}{r} 38 \\ 47 \\ +\,65 \\ \hline \end{array}$$

Name ____________________

Tiles in the bag: ____ Red ____ Green ____ Yellow ____ Blue

Bag ____

Color Tiles Picked

Red										
Green										
Yellow										
Blue										

Tiles in the bag: ______ Red ______ Green ______ Yellow ______ Blue

Bag ______

Color Tiles Picked

Red										
Green										
Yellow										
Blue										

Name ___________________________

Class Fact Practice 121A

Saxon *Math 2 (for use with* ***Lesson 121****)*

Set 24: Multiplying by 3

3×0	3×1	3×2	3×3	3×4
3×5	3×6	3×7	3×8	3×9
3×6	3×2	3×8	3×0	3×4
3×5	3×1	3×9	3×7	3×3
3×4	3×6	3×5	3×8	3×9

Name ______________________________

Fact Homework 121B

Saxon Math 2 (for use with Lesson 121)

Set 24: Multiplying by 3

1. Read the answers to someone.
2. Write the answers.
3. Ask someone to correct your paper. Corrected by ______________________

3×0	3×1	3×2	3×3	3×4
3×5	3×6	3×7	3×8	3×9
3×4	3×1	3×8	3×2	3×5
3×3	3×9	3×0	3×6	3×7
3×8	3×4	3×6	3×3	3×9

Name ____________________

***Saxon** Math 2 (for use with **Lesson 121**)*

A.

B.

C.

Name •

Draw a 3" line segment.

Date •——————————•

Measure this line segment using inches. ____ "

Guided Class Practice 121A

Saxon Math 2 (for use with Lesson 121)

1. On Monday the cafeteria served 254 hot lunches. On Tuesday they served 329 hot lunches. How many hot lunches did they serve on these two days altogether?

Number sentence ____________________

Answer ____________________

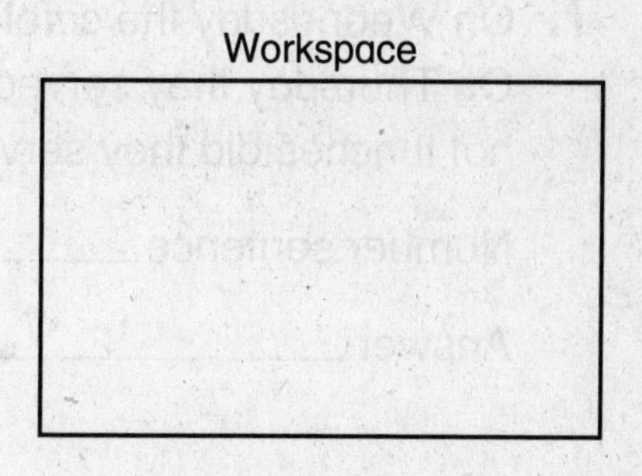

2. Twelve children chose vanilla ice cream, nine children chose chocolate ice cream, and five children chose strawberry ice cream.

Shade the graph to show the ice cream flavors the children chose.

How many more children chose vanilla than chocolate? ______

3. Label these arrays.

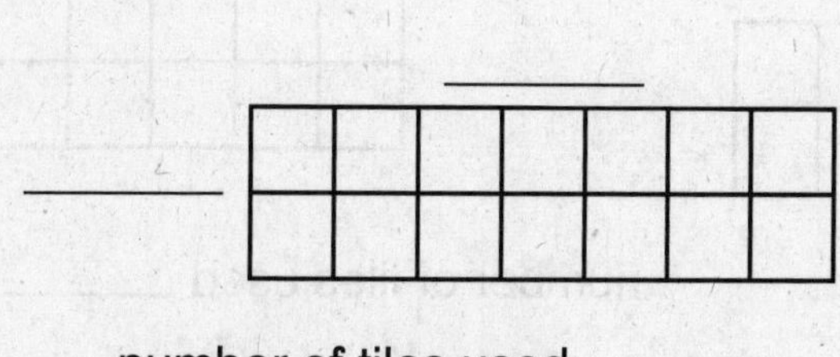

number of tiles used ______

______ × ______ array

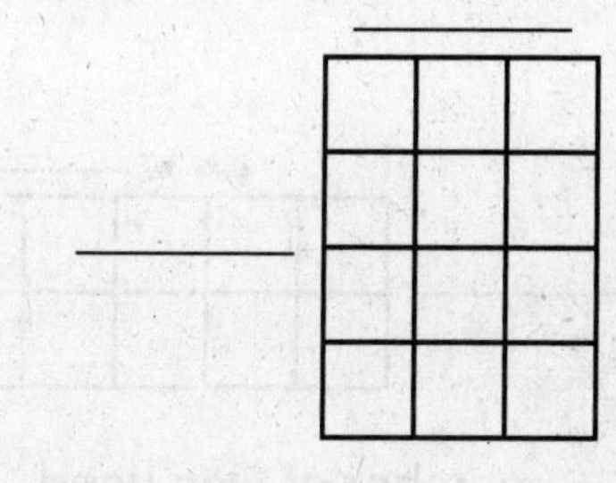

number of tiles used ______

______ × ______ array

4. Use a crayon to trace a pair of perpendicular line segments in Problem 3.

5. Find the answers.

725 − 416	\$3.75 − 1.92	287 + 15	\$6.29 + 2.81

Name ______________________

Date ______________________

Homework 121B

Saxon *Math 2 (for use with* ***Lesson 121****)*

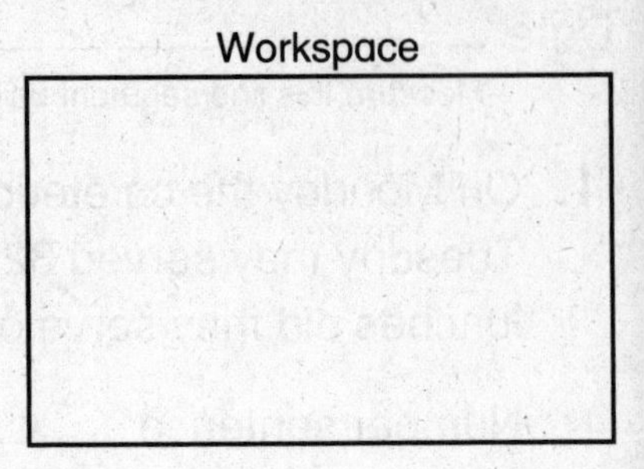

1. On Wednesday the cafeteria served 317 hot lunches. On Thursday they served 274 hot lunches. How many hot lunches did they serve on these two days altogether?

Number sentence ______________________

Answer ______________________

2. Seven children chose bananas, ten children chose apples, and eleven children chose oranges.

Shade the graph to show the fruits the children chose.

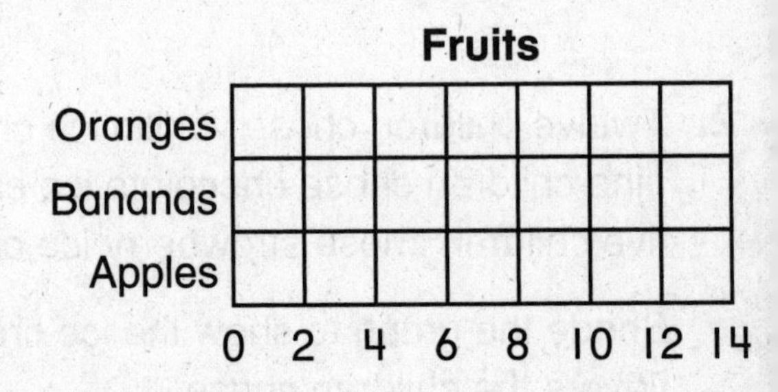

How many more children chose oranges than apples? ________

3. Label these arrays.

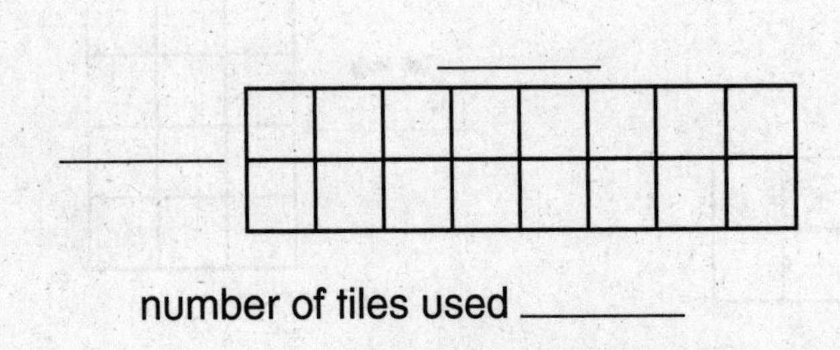

number of tiles used ________

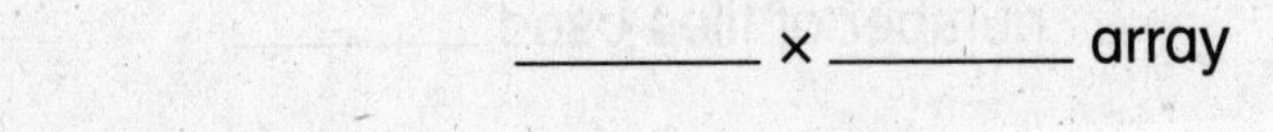

______ × ______ array

number of tiles used ________

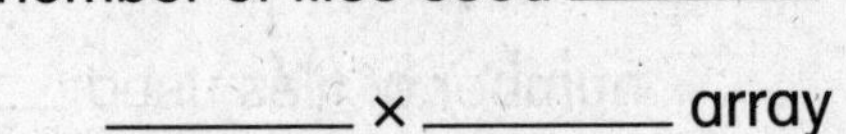

______ × ______ array

4. Use a crayon to trace a pair of perpendicular line segments in Problem 3.

5. Find the answers.

$$\begin{array}{r} 627 \\ -\ 292 \\ \hline \end{array} \qquad \begin{array}{r} \$7.54 \\ -\ 1.28 \\ \hline \end{array} \qquad \begin{array}{r} 636 \\ +\ 95 \\ \hline \end{array} \qquad \begin{array}{r} \$1.58 \\ +\ 6.27 \\ \hline \end{array}$$

Name ______________________

Fact Homework 122

Saxon Math 2 (for use with Lesson 122)

Set 24: Multiplying by 3

Corrected by ______

1. Fill in the products.

$0 \times 3 =$ $5 \times 3 =$

$1 \times 3 =$ $6 \times 3 =$

$2 \times 3 =$ $7 \times 3 =$

$3 \times 3 =$ $8 \times 3 =$

$4 \times 3 =$ $9 \times 3 =$

2. Match the problems to the answers.

2×3 •	• 9
7×3 •	• 18
3×3 •	• 6
6×3 •	• 21
9×3 •	• 3
1×3 •	• 27
4×3 •	• 24
8×3 •	• 15
0×3 •	• 12
5×3 •	• 0

3. Fill in the missing factors.

$\square \times 3 = 15$

$\square \times 3 = 0$

$\square \times 3 = 24$

$\square \times 3 = 9$

$\square \times 3 = 18$

Name

Draw a 7-cm line segment.

Date

Measure this line segment using centimeters. ______ cm

Guided Class Practice 122A

***Saxon** Math 2 (for use with **Lesson 122**)*

Workspace

1. The children in Mrs. Carlisle's class read 337 books, and the children in Mrs. Bueter's class read 284 books. How many books did the children in the two classes read altogether?

Number sentence ______________________

Answer ______________________

2. Write number sentences for these arrays.

✗✗✗✗✗✗✗
✗✗✗✗✗✗✗ ______________

○○○
○○○
○○○
○○○ ______________

3. Color $2\frac{3}{4}$ circles.

○ ○ ○

4. How many yellow tiles are in Bag A? ________

Which color has one more tile than green? ____________

Which two colors make up half of the tiles in Bag A?

blue and yellow   red and blue   green and blue

Color Tiles in Bag A

5 green
6 blue
3 yellow
2 red

5. I have 2 quarters, 1 dime, 3 nickels, and 1 penny. Draw the coins.

How much money do I have? ____________

6. Trace a pair of perpendicular line segments in Problem 5 using a crayon.

7. Find the answers.

62 − 48   563 + 194   494 − 277

−   +   −

Name ______________________

Date ______________________

Saxon Math 2 (for use with Lesson 122)

Workspace

1. The children in Mrs. Rosenberg's class read 352 books, and the children in Mrs. Limosani's class read 283 books. How many books did the children in the two classes read altogether?

Number sentence ______________________

Answer ______________________

2. Write number sentences for these arrays.

△ △ △ △ △
△ △ △ △ △ ______________

○○○○
○○○○
○○○○ ______________

3. Color $3\frac{3}{4}$ squares.

4. How many yellow tiles are in Bag B? ________

Which color has one more tile than green? ____________

Which two colors make up half of the tiles in Bag B?

blue and yellow   red and blue   green and blue

Color Tiles in Bag B

4 yellow
7 red
6 blue
3 green

5. I have 1 quarter, 3 dimes, 1 nickel, and 9 pennies. Draw the coins.

How much money do I have? ____________

6. Trace a pair of perpendicular line segments in Problem 5 using a crayon.

7. Find the answers.

73 – 48   475 + 370   817 – 430

Name ______________________________

Class Fact Practice 123A

Saxon *Math 2 (for use with* ***Lesson 123****)*

Set 24: Multiplying by 3

3×2	3×5	3×8	3×1	3×6
3×3	3×9	3×0	3×7	3×4
3×8	3×3	3×2	3×5	3×0
3×6	3×1	3×9	3×4	3×7
3×3	3×5	3×9	3×6	3×2

Name ______________________________

Saxon Math 2 (for use with Lesson 123)

Set 24: Multiplying by 3

1. Read the answers to someone.
2. Write the answers.
3. Ask someone to correct your paper. Corrected by ____________________

3×6	3×2	3×5	3×0	3×9
3×4	3×1	3×7	3×3	3×8
3×5	3×4	3×0	3×7	3×3
3×1	3×9	3×2	3×6	3×8
3×4	3×6	3×2	3×9	3×3

Name ______________________

Lesson Worksheet 123A

Saxon Math 2 (for use with Lesson 123)

1.

:

____

2.

:

____

____

3.

:

____

4.

:

____

____

5.

12
1
2
3
4
5
6
7
8
9
10
11

:

6.

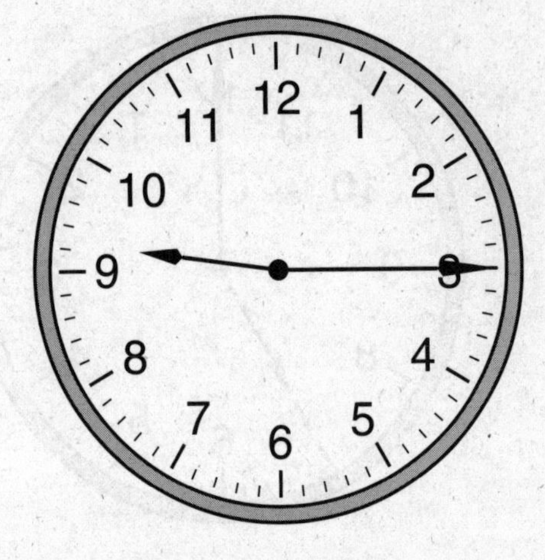
12
1
2
3
4
5
6
7
8
9
10
11

:

7.

12
1
2
3
4
5
6
7
8
9
10
11

:

8.

12
1
2
3
4
5
6
7
8
9
10
11

:

Name •

Draw a 3" line segment.

Date •———————————————•

Measure this line segment using inches. ______"

Guided Class Practice 123A

Saxon *Math 2 (for use with* ***Lesson 123****)*

1. Five children in Mrs. O'Neill's class made books. Each book had 3 pictures. How many pictures did the children draw altogether?

Number sentence ____________________

Answer ____________________

2. Circle the time shown on the clock.

quarter past 4 quarter to 5

quarter to 4 quarter past 5

3. Fill in the missing numbers in these number patterns.

25, 50, 75, ______, ______, ______, ______, ______, ______

______, ______, ______, 44, 54, 64, ______, ______, ______

4. Lauren has 6 red socks, 12 white socks, and 4 black socks in her drawer. If she takes out one sock without looking, which of these colors is she most likely to get?

____

5. Circle the number sentence for this array.

$3 + 6 = 9$ $15 + 3 = 18$

$3 \times 6 = 18$ $6 \times 6 = 36$

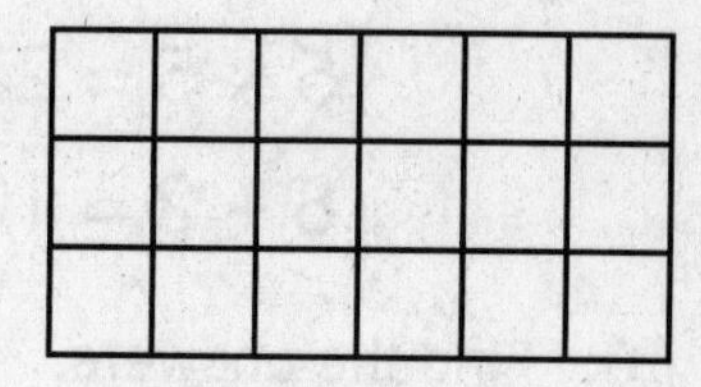

6. Find the answers.

$82 - 43$ $91 + 72 + 38$ $\$4.72 + \$2.19$

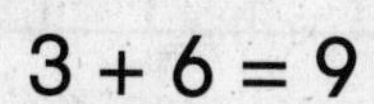

−

______

+

______

+

______

Name ______________________

Date ______________________

Saxon Math 2 (for use with **Lesson 123**)

1. Four children in Mrs. Sheehan's class made books. Each book had 5 pictures. How many pictures did the children draw altogether?

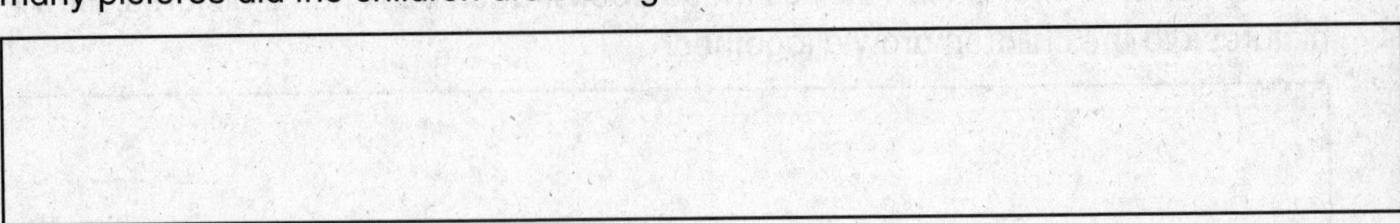

Number sentence ______________________

Answer ______________________

2. Circle the time shown on the clock.

quarter past 4 quarter past 5

quarter to 4 quarter to 5

3. Fill in the missing numbers in these number patterns.

100, 125, 150, ____, ____, ____, ____, ____, ____

____, ____, ____, 56, 66, 76, ____, ____, ____

4. Steven has 6 white socks, 10 black socks, and 4 blue socks in his drawer. If he takes out one sock without looking, which of these colors is he most likely to get?

____

5. Circle the number sentence for this array.

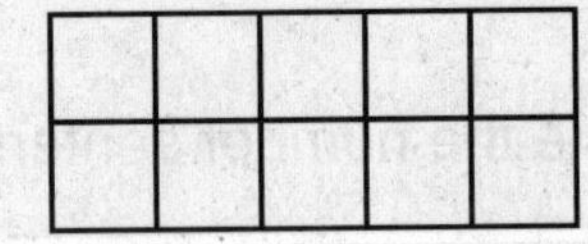

$5 \times 5 = 25$ $2 \times 5 = 10$

$8 + 2 = 10$ $5 + 2 = 7$

6. Find the answers.

$86 - 27$ $35 + 29 + 31$ $\$2.96 + \$3.72$

–

+

+

Name ______________________________

Class Fact Practice 124A

***Saxon** Math 2 (for use with **Lesson 124**)*

Set 24: Multiplying by 3

3×3	3×9	3×4	3×1	3×6
3×8	3×6	3×5	3×7	3×2
3×6	3×4	3×9	3×1	3×8
3×3	3×7	3×2	3×5	3×0
3×7	3×1	3×4	3×8	3×3

Name ______________________

Set 24: Multiplying by 3

1. Read the answers to someone.
2. Write the answers.
3. Ask someone to correct your paper. Corrected by ______________________

3×7	3×0	3×5	3×9	3×1
3×4	3×8	3×2	3×6	3×3
3×6	3×4	3×1	3×8	3×2
3×0	3×7	3×5	3×3	3×9
3×4	3×8	3×2	3×7	3×3

Name ______________________________

Translation—Slide

Rotation—Turn

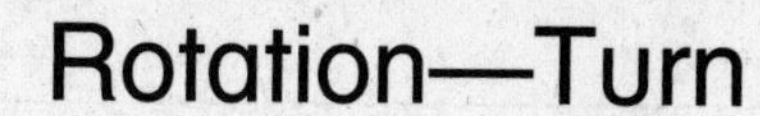

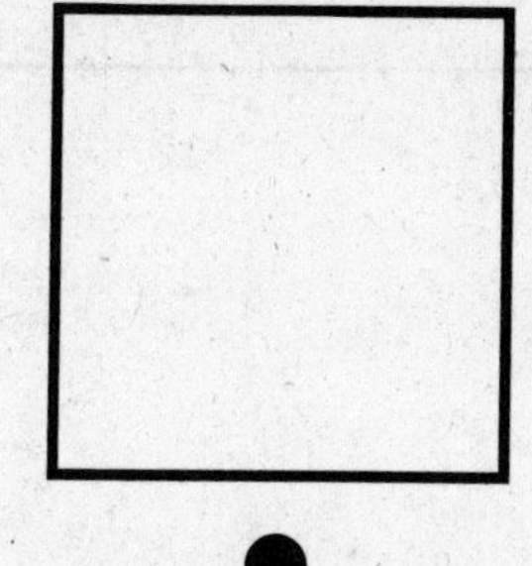

Reflection—Flip

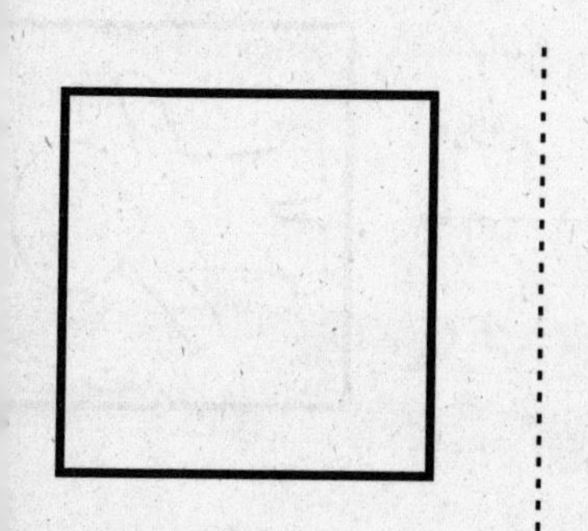

Fat Cat

Cookie Man

Name ___

1. Start Finish

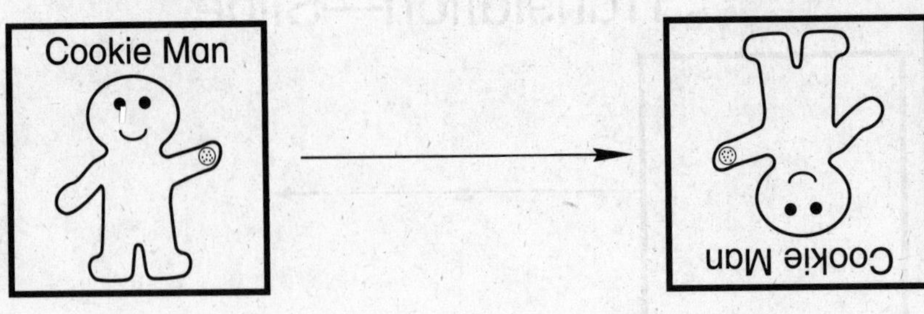

___

2. Start Finish

Fat Cat

___

3. Start Finish

___

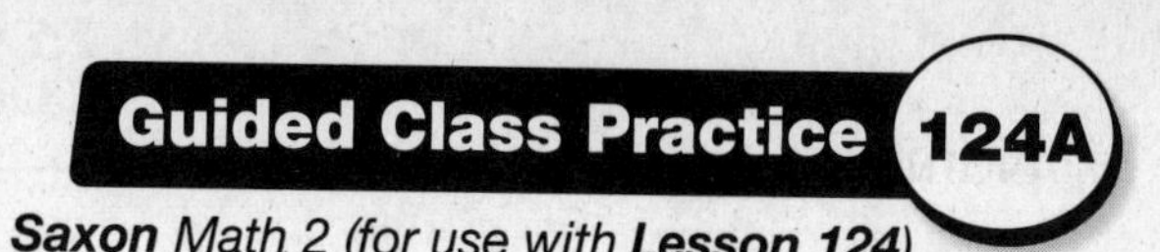

Name
Draw a 7-cm line segment.

Saxon Math 2 (for use with Lesson 124)

Date

Measure this line segment using centimeters. ______ cm

Workspace

1. Michael has 269 pennies and Willie has 185 pennies.
How many pennies do the two boys have altogether?

Number sentence ______________________

Answer ______________________

2. Write a mixed number to show how many circles are shaded.

3. Draw a triangle that has a right angle in the box.

4. Label this array.

Write a number sentence for the array. ______________________

5. Draw a pictograph to show how many tiles of each color are in Bag A.

Bag A

Color	Tiles
Red	6
Blue	3
Yellow	12

Tiles in Bag A

Red	
Blue	
Yellow	

☐ = 2 tiles

If you take one tile out of the bag without looking, which of these colors are you least likely to get? ______________________

Name a color it will be impossible to get. ______________________

6. Find the answers.

$$\begin{array}{r} 556 \\ +\ 84 \\ \hline \end{array} \qquad \begin{array}{r} \$5.21 \\ +\ 3.79 \\ \hline \end{array} \qquad \begin{array}{r} 380 \\ -\ 142 \\ \hline \end{array} \qquad \begin{array}{r} \$6.90 \\ -\ 2.36 \\ \hline \end{array}$$

Name ______________________

Date ______________________

Workspace

1. Flavia has 193 pennies and Carmela has 227 pennies.
How many pennies do the two girls have altogether?

Number sentence ______________________

Answer ______________________

2. Write a mixed number to show how many squares are shaded.

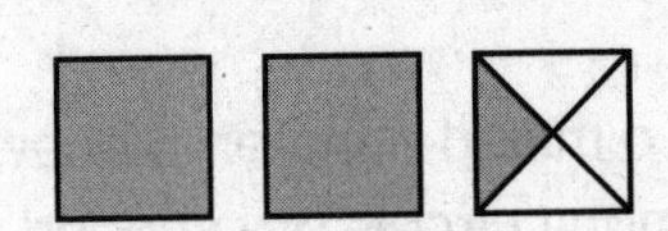

3. Draw a shape that has 4 right angles in the box.

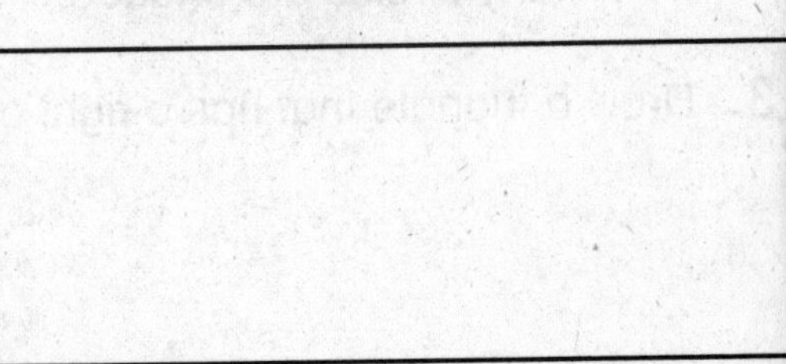

4. Label this array.

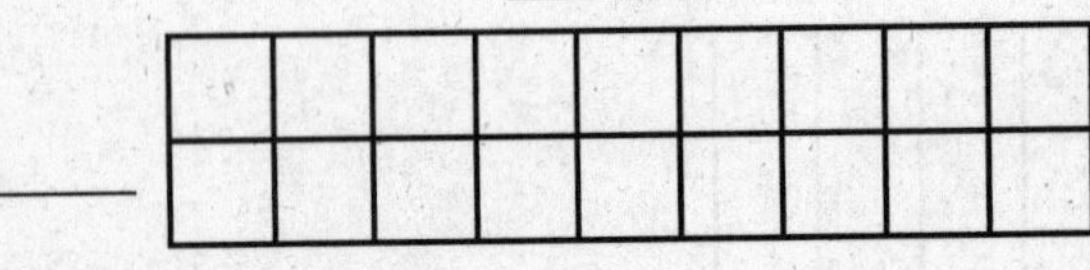

Write a number sentence for the array. ______________________

5. Draw a pictograph to show how many tiles of each color are in Bag B.

Bag B

Color	Tiles
Red	2
Blue	9
Yellow	5

Tiles in Bag B

Red	
Blue	
Yellow	

☐ = 2 tiles

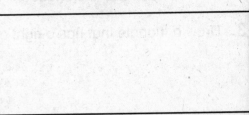

If you take one tile out of the bag without looking, which of these colors are you most likely to get? ______________________

Name a color it will be impossible to get. ______________________

6. Find the answers.

$$\begin{array}{r} 397 \\ +\ 63 \\ \hline \end{array} \qquad \begin{array}{r} \$3.73 \\ +\ 2.27 \\ \hline \end{array} \qquad \begin{array}{r} 520 \\ -\ 218 \\ \hline \end{array} \qquad \begin{array}{r} \$7.6 \\ -\ 4.2 \\ \hline \end{array}$$

Name ______________________ Score ______

Fact Assessment 24-1

Saxon Math 2 (for use with Lesson 125-1)

Set 24: Multiplying by 3

3×3	3×5	3×0	3×7	3×2
3×9	3×1	3×6	3×4	3×8
3×2	3×7	3×4	3×9	3×0
3×5	3×1	3×8	3×3	3×6
3×4	3×0	3×2	3×6	3×5

Name ______________________________

Fact Homework 125A

***Saxon** Math 2 (for use with **Lesson 125-1**)*

Set 24: Multiplying by 3

Pretend you are the teacher.
Correct this paper.
If the answer is incorrect, write the correct answer next to the problem.

$3 \times 3 = 9$	$3 \times 5 = 15$	$3 \times 0 = 3$	$3 \times 7 = 18$	$3 \times 2 = 6$
$3 \times 9 = 27$	$3 \times 1 = 3$	$3 \times 6 = 18$	$3 \times 4 = 12$	$3 \times 8 = 24$
$3 \times 2 = 6$	$3 \times 7 = 21$	$3 \times 4 = 15$	$3 \times 9 = 27$	$3 \times 0 = 0$
$3 \times 5 = 15$	$3 \times 1 = 3$	$3 \times 8 = 26$	$3 \times 3 = 9$	$3 \times 6 = 21$
$3 \times 4 = 12$	$3 \times 0 = 0$	$3 \times 2 = 6$	$3 \times 6 = 18$	$3 \times 5 = 15$

Name ____________________

S100: 100 Subtraction Facts   Corrected by ________

1	7 − 1	10 − 4	9 − 0	16 − 9	5 − 4	12 − 6	9 − 7	11 − 3	8 − 2	6 − 6
2	9 − 4	6 − 3	11 − 6	10 − 2	6 − 1	12 − 8	2 − 0	9 − 3	7 − 2	6 − 5
3	5 − 5	11 − 4	4 − 2	15 − 9	8 − 0	10 − 6	14 − 5	9 − 9	4 − 0	12 − 7
4	9 − 5	17 − 9	8 − 4	13 − 8	9 − 2	11 − 5	15 − 6	5 − 1	8 − 5	16 − 8
5	8 − 6	11 − 7	1 − 0	7 − 3	9 − 6	4 − 3	17 − 8	10 − 5	12 − 4	13 − 7
6	8 − 3	16 − 7	10 − 3	4 − 1	6 − 2	13 − 5	7 − 0	14 − 9	11 − 2	10 − 8
7	13 − 9	10 − 7	18 − 9	14 − 6	1 − 1	12 − 3	7 − 5	2 − 1	11 − 8	7 − 7
8	2 − 2	12 − 5	3 − 1	15 − 7	10 − 1	6 − 0	13 − 4	5 − 2	9 − 8	3 − 0
9	11 − 9	7 − 6	13 − 6	3 − 3	14 − 8	9 − 1	6 − 4	12 − 9	7 − 4	8 − 7
10	4 − 4	15 − 8	3 − 2	5 − 0	5 − 3	8 − 8	14 − 7	10 − 9	0 − 0	8 − 1

Name ______________________________

Lesson Worksheet 125-1A

Saxon *Math 2 (for use with* ***Lesson 125-1****)*

1. Fill in the products.

$0 \times 4 =$ $\quad$ $5 \times 4 =$

$1 \times 4 =$ $\quad$ $6 \times 4 =$

$2 \times 4 =$ $\quad$ $7 \times 4 =$

$3 \times 4 =$ $\quad$ $8 \times 4 =$

$4 \times 4 =$ $\quad$ $9 \times 4 =$

2. Match the problems to the answers.

4×4 •	• 32
0×4 •	• 0
8×4 •	• 16
6×4 •	• 4
1×4 •	• 20
9×4 •	• 24
5×4 •	• 28
2×4 •	• 36
7×4 •	• 12
3×4 •	• 8

3. Fill in the missing factors.

$\square \times 4 = 28$

$\square \times 4 = 0$

$\square \times 4 = 16$

$\square \times 4 = 36$

$\square \times 4 = 4$

Name ______________________

Lesson Worksheet 125-

Saxon Math 2 (first used in Lesson 125-1)

Name

Draw a $2\frac{1}{2}$" line segment.

Saxon Math 2 (for use with Lesson 125-1)

Date

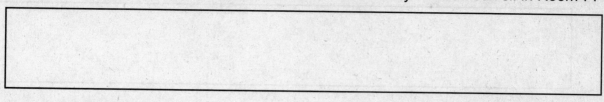

Measure this line segment using inches. ______"

1. Three children can sit at each table in Room 7. There are ten tables in the room. Draw a picture to show the tables and chairs in Room 7. How many children can sit in Room 7?

Number sentence ______________________

Answer ______________________

2. Label this array. Write a number sentence for the array.

____

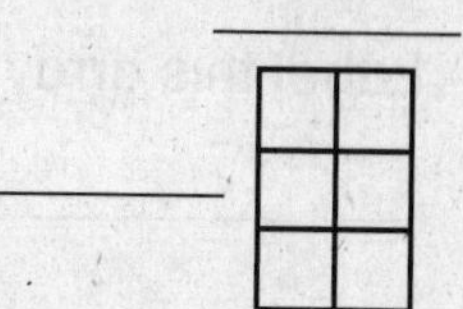

3. Circle the best number sentence to use to estimate the sum of 63 and 29.

$60 + 20 = 80$ $60 + 30 = 90$ $70 + 20 = 90$ $70 + 30 = 100$

4. Sandy's family ate $2\frac{1}{2}$ cakes. Color the cakes to show how much cake they ate. How much cake is left?

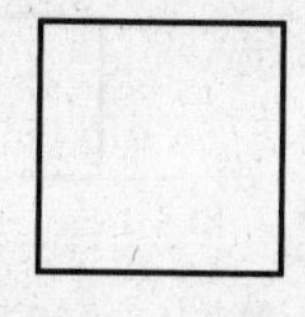

5. Circle the clock that shows quarter to 6.

6. Find the answers.

$$\begin{array}{r} 217 \\ +\ 394 \\ \hline \end{array} \qquad \begin{array}{r} 621 \\ -\ 73 \\ \hline \end{array} \qquad \begin{array}{r} 58 \\ 9 \\ +\ 33 \\ \hline \end{array}$$

Name ______________________________

Date ______________________________

1. Mrs. Wagoner has 5 games for the children to use during recess. Four children can play each game. Draw a picture to show the games and children. How many children can play the games?

Number sentence ______________________________

Answer ______________________________

2. Label this array. Write a number sentence for the array.

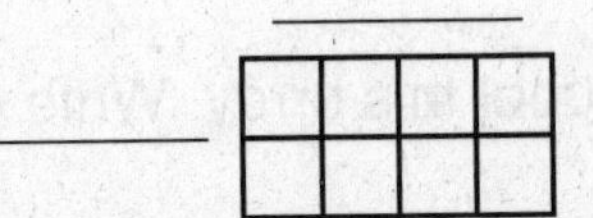

____

3. Circle the best number sentence to use to estimate the sum of 28 and 49.

20 + 40 = 60   20 + 50 = 70   30 + 40 = 70   30 + 50 = 80

4. Carol's family ate $1\frac{1}{2}$ cakes. Color the cakes to show how much cake they ate. How much cake is left?

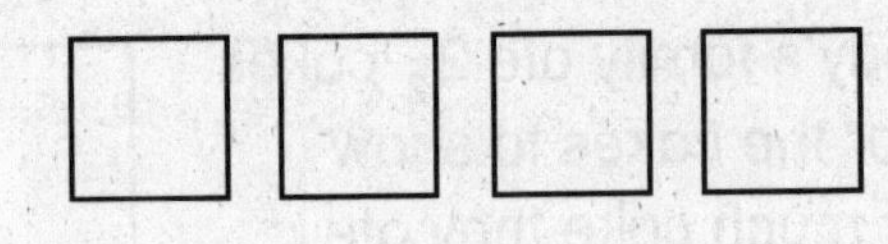

5. Circle the clock that shows quarter to 5.

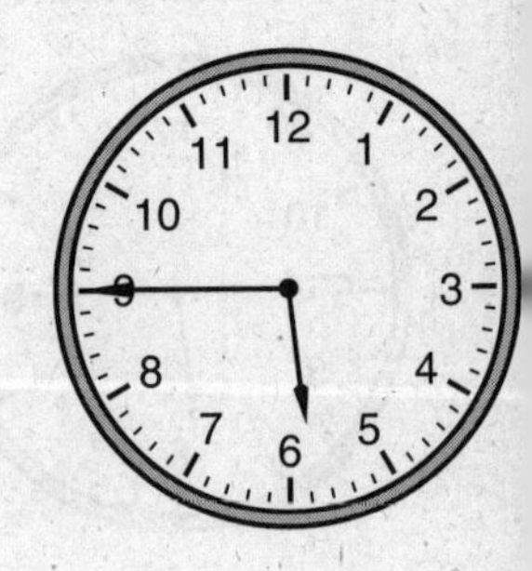

6. Find the answers.

$$\begin{array}{r} 429 \\ +\ 372 \\ \hline \end{array} \qquad \begin{array}{r} 841 \\ -\ 69 \\ \hline \end{array} \qquad \begin{array}{r} 42 \\ 8 \\ +\ 71 \\ \hline \end{array}$$

Name ______________________ Score ______

Fact Assessment 24-2

***Saxon** Math 2 (for use with **Lesson 125-2**)*

S100: 100 Subtraction Facts

1	7 − 1	10 − 4	9 − 0	16 − 9	5 − 4	12 − 6	9 − 7	11 − 3	8 − 2	6 − 6
2	9 − 4	6 − 3	11 − 6	10 − 2	6 − 1	12 − 8	2 − 0	9 − 3	7 − 2	6 − 5
3	5 − 5	11 − 4	4 − 2	15 − 9	8 − 0	10 − 6	14 − 5	9 − 9	4 − 0	12 − 7
4	9 − 5	17 − 9	8 − 4	13 − 8	9 − 2	11 − 5	15 − 6	5 − 1	8 − 5	16 − 8
5	8 − 6	11 − 7	1 − 0	7 − 3	9 − 6	4 − 3	17 − 8	10 − 5	12 − 4	13 − 7
6	8 − 3	16 − 7	10 − 3	4 − 1	6 − 2	13 − 5	7 − 0	14 − 9	11 − 2	10 − 8
7	13 − 9	10 − 7	18 − 9	14 − 6	1 − 1	12 − 3	7 − 5	2 − 1	11 − 8	7 − 7
8	2 − 2	12 − 5	3 − 1	15 − 7	10 − 1	6 − 0	13 − 4	5 − 2	9 − 8	3 − 0
9	11 − 9	7 − 6	13 − 6	3 − 3	14 − 8	9 − 1	6 − 4	12 − 9	7 − 4	8 − 7
10	4 − 4	15 − 8	3 − 2	5 − 0	5 − 3	8 − 8	14 − 7	10 − 9	0 − 0	8 − 1

Name ______________________

Date ______________________

1. There are 257 children at Stiles School. There are 623 children at Savin Rock School. How many children are there at the two schools altogether?

Number sentence ______________________

Answer ______________________

Workspace

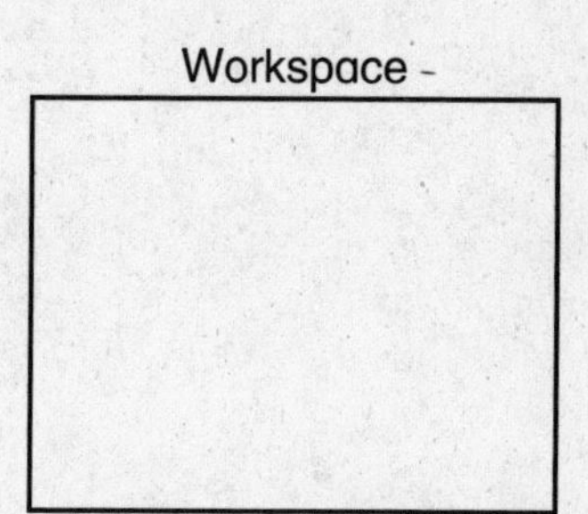

2. How many children are in Room 8? ________

How many children are in Room 9? ________

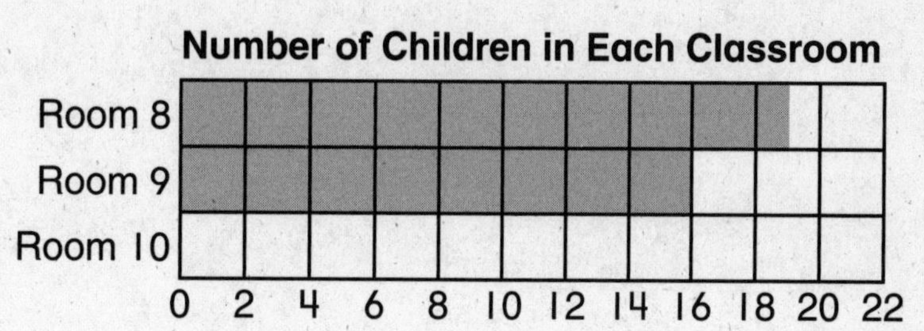

Shade the graph to show that there are 20 children in Room 10.

How many more children are in Room 10 than in Room 9? ________

3. Circle the shapes that have a right angle.

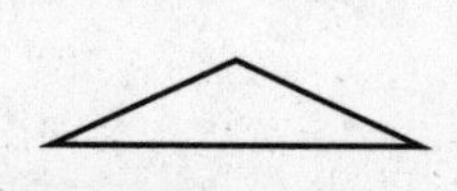

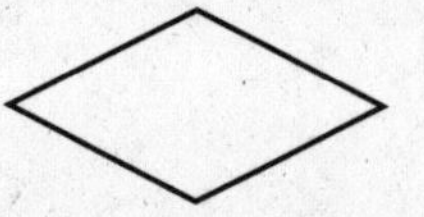
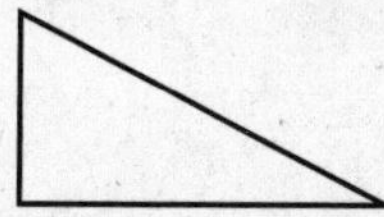

4. Color $3\frac{1}{4}$ circles.

5. Fill in the missing numbers in the number patterns.

70, 75, 80, ______, ______, ______, ______, ______, ______

______, ______, ______, 58, 68, 78, ______, ______, ______

6. Find the answers.

$$\begin{array}{r} 65 \\ -\ 27 \\ \hline \end{array} \qquad \begin{array}{r} 80 \\ -\ 21 \\ \hline \end{array} \qquad \begin{array}{r} 325 \\ +\ 267 \\ \hline \end{array} \qquad \begin{array}{r} 549 \\ +\ 190 \\ \hline \end{array}$$

Name ______________________

Question ______________________

Choices 1) ______________________

2) ______________________

3) ______________________

4) ______________________

The title of the graph will be:

____

Choices	Tally of Votes
1) ______	
2) ______	
3) ______	
4) ______	

If any choice receives more than 7 votes, number your graph by 2's.

Name ________________________________

Lesson Worksheet 125-2B

***Saxon** Math 2 (for use with **Lesson 125-2**)*

Title ____

________________ ________________ ________________ ________________

Class surveyed: Grade ______ **Teacher** ________________________

Name ______________________

Class Fact Practice 126A

***Saxon** Math 2 (for use with **Lesson 126**)*

Set 25: Multiplying by 4

4×0 | 4×1 | 4×2 | 4×3 | 4×4

4×5 | 4×6 | 4×7 | 4×8 | 4×9

4×6 | 4×2 | 4×8 | 4×0 | 4×4

4×5 | 4×1 | 4×9 | 4×7 | 4×3

4×4 | 4×6 | 4×5 | 4×8 | 4×9

Name ______________________________

Saxon Math 2 (for use with **Lesson 126**)

Set 25: Multiplying by 4

1. Read the answers to someone.
2. Write the answers.
3. Ask someone to correct your paper. Corrected by ____________________

4×0	4×1	4×2	4×3	4×4
4×5	4×6	4×7	4×8	4×9
4×4	4×1	4×8	4×2	4×5
4×3	4×9	4×0	4×6	4×7
4×8	4×4	4×6	4×3	4×9

Name ______________________________

4	•	•	•	•	•
3	•	•	•	•	•
2	•	•	•	•	•
1	•	•	•	•	•
0	•	•	•	•	•
	0	1	2	3	4

☺ (,) (4, 3) 

☆ (,) 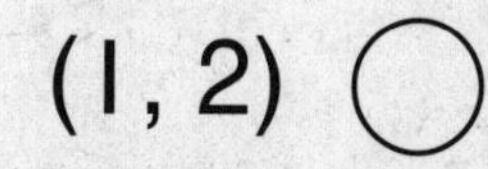 (1, 2) ○

♡ (,) (2, 0) □

Name •
Draw an 8-cm line segment.

Date •———————————•
Measure this line segment using centimeters. ________ cm

Guided Class Practice 126A

***Saxon** Math 2 (for use with **Lesson 126**)*

Workspace

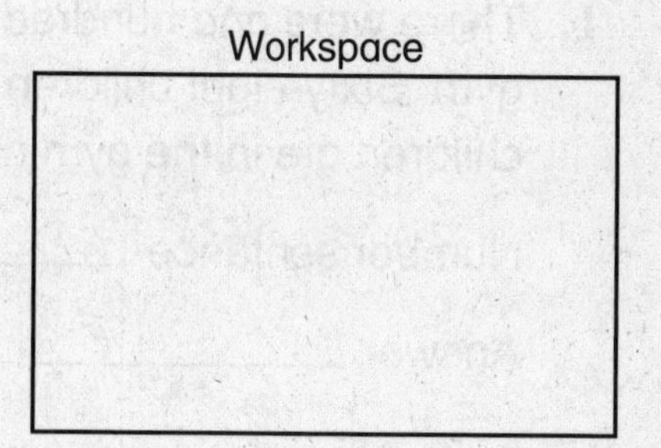

1. There were one hundred twenty-six children in the gym. Eighty-seven children joined them. How many children are in the gym now?

Number sentence ______________________

Answer ______________________

2. Use the Venn diagram to answer the questions.

Which letters on the graph have parallel line segments, but not perpendicular line segments? ____________

Which letters have perpendicular line segments? ____________

Which letters have both parallel and perpendicular line segments? ____________

Letters

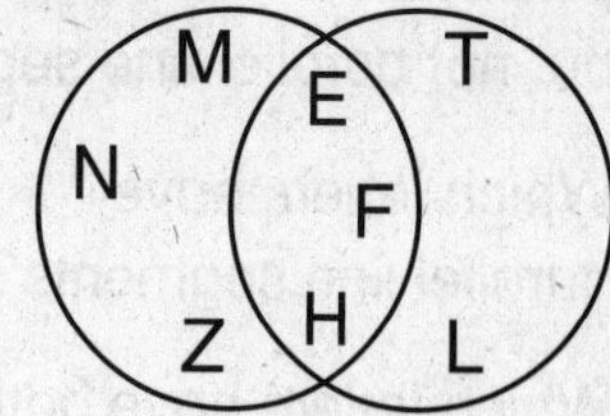

3. Put a red dot at (3, 0).

Put a blue dot at (2, 4).

4 • • • • •
3 • • • • •
2 • • • • •
1 • • • • •
0 • • • • •
0 1 2 3 4

4. Circle what the 5 will look like when you slide and flip it.

 5  

5. Fill in the correct symbol (+, –, or ×).

$5 + 9 = 15 \bigcirc 1$ ________

$2 = 4 \bigcirc 2$

$2 \times 10 = 4 \bigcirc 5$ ________

6. Find the answers.

470 – 142

–
____

$5.63 + $1.88

+
____

42	29
38	59
+ 17	+ 32

Name ______________________________

Date ______________________________

Workspace

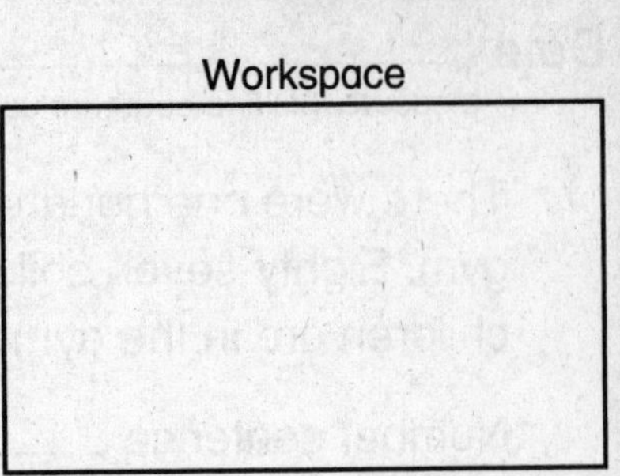

1. There were one hundred forty-three children in the gym. Sixty-eight children joined them. How many children are in the gym now?

Number sentence ______________________________

Answer ______________________________

2. Use the Venn diagram to answer the questions.

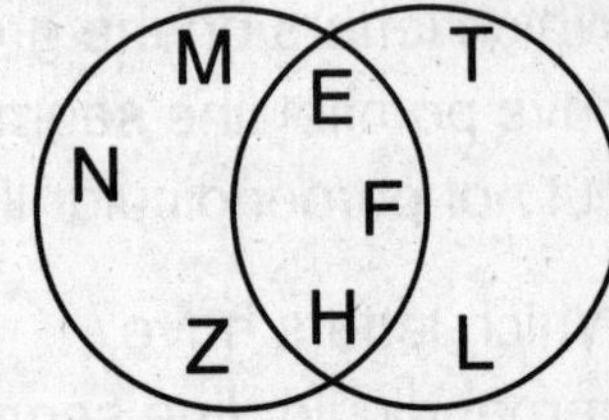

Which letters on the graph have perpendicular line segments, but not parallel line segments? ____________

Which letters have parallel line segments? ____________

Which letters have both parallel and perpendicular line segments? ____________

3. Put a red dot at (3, 4).

Put a blue dot at (1, 2).

4 • • • • •
3 • • • • •
2 • • • • •
1 • • • • •
0 • • • • •
0 1 2 3 4

4. Circle what the 3 will look like when you slide and flip it.

5. Fill in the correct symbol (+, −, or ×).

16 − 9 = 4 ◯ 3   3 = 3 ◯ 1   3 × 5 = 10 ◯ 5

6. Find the answers.

618 − 254 $4.52 + $1.79

$$\begin{array}{r} 38 \\ 52 \\ +\ 76 \\ \hline \end{array} \qquad \begin{array}{r} 16 \\ 57 \\ +\ 62 \\ \hline \end{array}$$

Name ______________________________

Saxon Math 2 (for use with **Lesson 127**)

Set 25: Multiplying by 4 Corrected by ________

1. Fill in the products.

$0 \times 4 =$ $5 \times 4 =$

$1 \times 4 =$ $6 \times 4 =$

$2 \times 4 =$ $7 \times 4 =$

$3 \times 4 =$ $8 \times 4 =$

$4 \times 4 =$ $9 \times 4 =$

2. Match the problem to the answers.

2×4 •	• 24
6×4 •	• 8
3×4 •	• 32
1×4 •	• 20
8×4 •	• 12
5×4 •	• 4
0×4 •	• 16
7×4 •	• 36
4×4 •	• 0
9×4 •	• 28

3. Fill in the missing factors.

$\square \times 4 = 12$

$\square \times 4 = 32$

$\square \times 4 = 20$

$\square \times 4 = 8$

$\square \times 4 = 24$

Name ______________________

1. 

____

Change from $1.00

2. 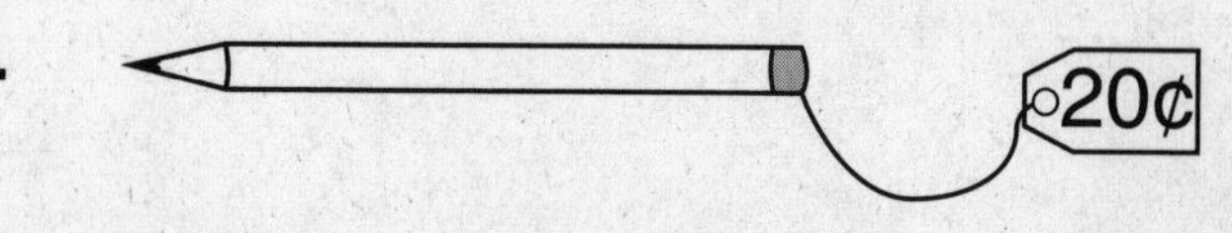

____

Change from $1.00

3. 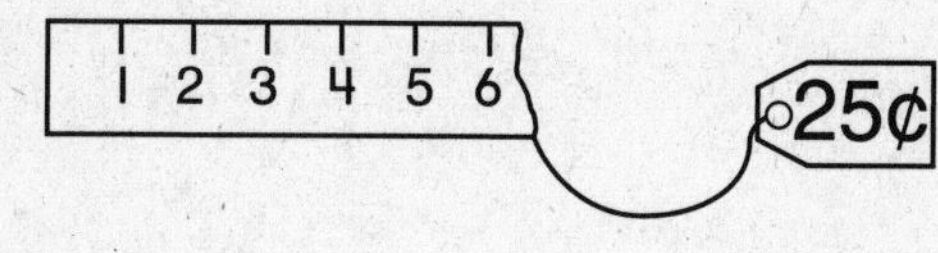

____

Change from $1.00

4. 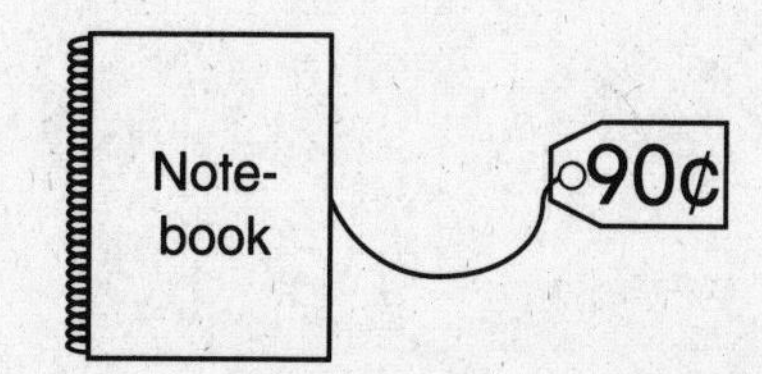

____

Change from $1.00

5. 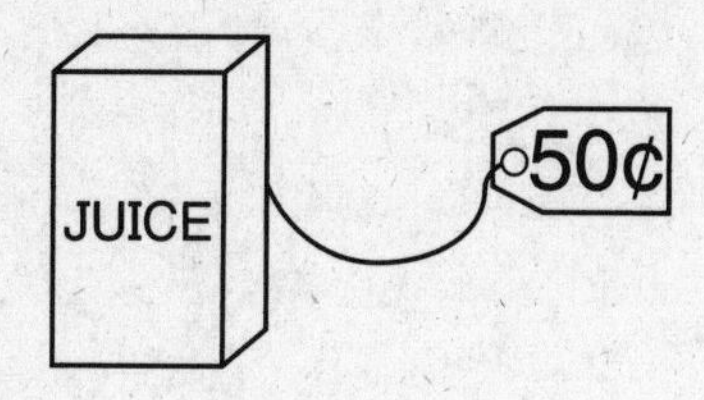

____

Change from $1.00

6. 

____

Change from $1.00

Name •
Draw a $3\frac{1}{2}$" line segment.

Guided Class Practice 127A

Saxon Math 2 (for use with Lesson 127)

Date •————————•
Measure this line segment using inches. ________"

1. Each child at the party ate 5 cookies. There were six children at the party. Draw a picture and write a number sentence to show how many cookies the children ate altogether.

Number sentence ______________________________

Answer ____________________

2. Circle the time shown on the clock.

quarter past 8   quarter to 8
quarter past 9   quarter to 9

3. The cost of a notebook is 75¢.
How much change will you receive from $1.00? ____________

4. Circle what the F will look like when you slide and flip it.

5. Circle the best number sentence to use to estimate the sum of 68 and 23.

60 + 20 = 80   60 + 30 = 90   70 + 20 = 90   70 + 30 = 100

6. If you take one marble out of the bag, which of these colors are you most likely to get? ________

Which of these colors are you least likely to get? ________________

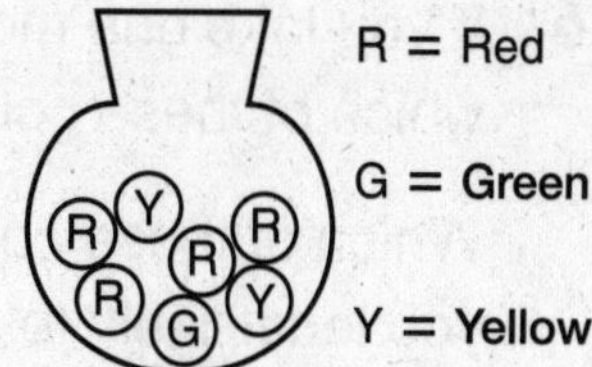

7. Find the answers.

486	359	$4.17	$6.18
− 79	+ 78	+ 3.95	− 0.74

Name ______________________

Date ______________________

Saxon Math 2 (for use with Lesson 127)

1. Each child at the party drank 3 cups of juice. There were seven children at the party. Draw a picture and write a number sentence to show how many cups of juice the children drank altogether.

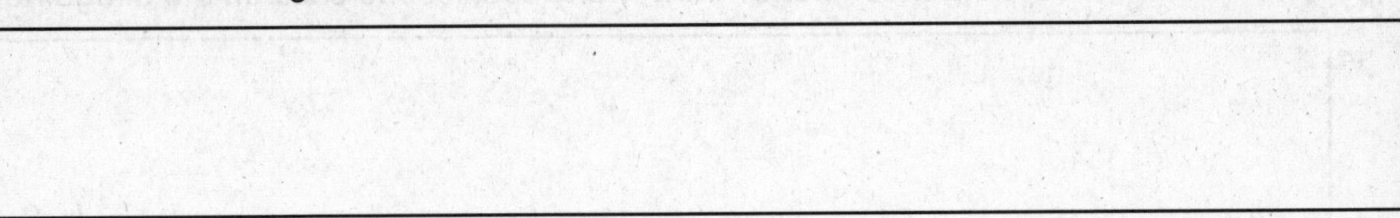

Number sentence ______________________

Answer ______________________

2. Circle the time shown on the clock.

quarter past 8 quarter to 8

quarter past 9 quarter to 9

3. The cost of a marker is 20¢.
How much change will you receive from $1.00? ______________

4. Circle what the L will look like when you slide and flip it.

5. Circle the best number sentence to use to estimate the sum of 31 and 48.

30 + 40 = 70 30 + 50 = 80 40 + 40 = 80 40 + 50 = 90

6. If you take one marble out of the bag, which of these colors are you most likely to get? ______________

Which of these colors are you least likely to get? ______________

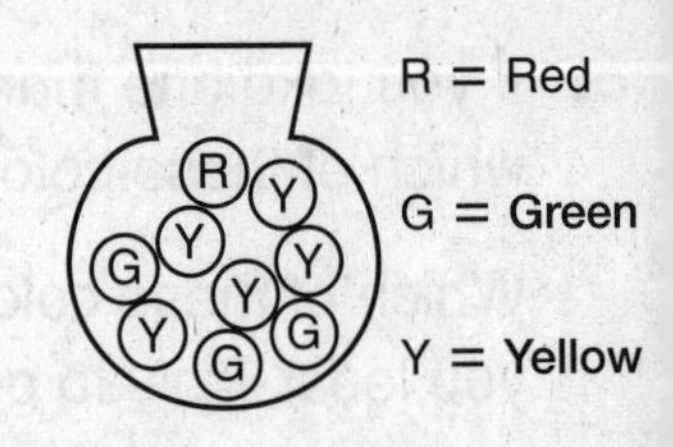

7. Find the answers.

671	529	$2.49	$3.57
− 34	+ 93	+ 3.72	− 2.84

Name ______________________________

Class Fact Practice 128A

Saxon *Math 2 (for use with* ***Lesson 128****)*

Set 25: Multiplying by 4

4×2	4×5	4×8	4×1	4×6
4×3	4×9	4×0	4×7	4×4
4×8	4×3	4×2	4×5	4×0
4×6	4×1	4×9	4×4	4×7
4×3	4×5	4×9	4×6	4×2

Name ____________________

Set 25: Multiplying by 4

1. Read the answers to someone.
2. Write the answers.
3. Ask someone to correct your paper. Corrected by ____________________

4×6	4×2	4×5	4×0	4×9
4×4	4×1	4×7	4×3	4×8
4×5	4×4	4×0	4×7	4×3
4×1	4×9	4×2	4×6	4×8
4×4	4×6	4×2	4×9	4×3

Name

Draw a 7-cm line segment.

Date

Measure this line segment using centimeters. ______ cm

Guided Class Practice 128A

***Saxon** Math 2 (for use with **Lesson 128**)*

1. The children in Mrs. Watkin's class collected 258 cans for recycling. They gave 175 cans to the custodian. How many cans do they have left?

Workspace

Number sentence ______________________

Answer ______________________

2. There are 3 orange, 3 green, 1 blue, and 7 yellow tiles in a bag. If you pick a tile without looking, which color will you most likely pick? __________

Why? ______________________

Which colors will you be equally likely to pick? __________

3. Write a number sentence for this array.

____

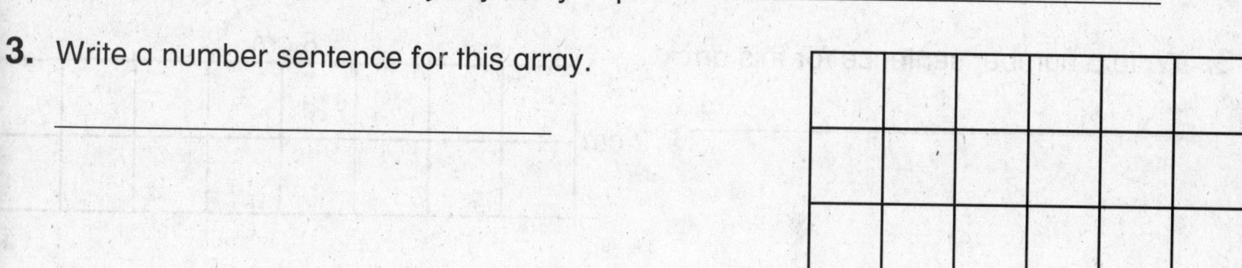

4. Measure the sides of the rectangle in Problem 3 using centimeters.

What is the perimeter of the rectangle? __________

5. Write a mixed number to show how many squares are shaded.

6. Put a red dot at (1, 3).
Put a blue dot at (2, 0).

7. Find the answers.

$6 \div 2 =$ ______ $4 \div 2 =$ ______

$18 \div 2 =$ ______ $12 \div 2 =$ ______

$$\begin{array}{r} 717 \\ +\ 208 \\ \hline \end{array} \qquad \begin{array}{r} 490 \\ -\ 146 \\ \hline \end{array}$$

Name ______________________

Date ______________________

Saxon Math 2 (for use with Lesson 128)

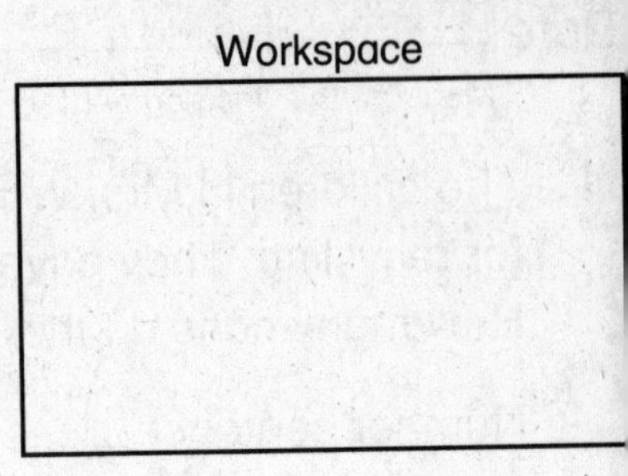

1. The children in Mrs. Burr's class collected 427 cans for recycling. They gave 125 cans to the custodian. How many cans do they have left?

Number sentence ______________________

Answer ______________________

2. There are 1 blue, 5 red, 4 orange, and 1 green tiles in a bag. If you pick a tile without looking, which color will you most likely pick? ______________

Why? ______________________

Which colors will you be equally likely to pick? ______________

3. Write a number sentence for this array.

____

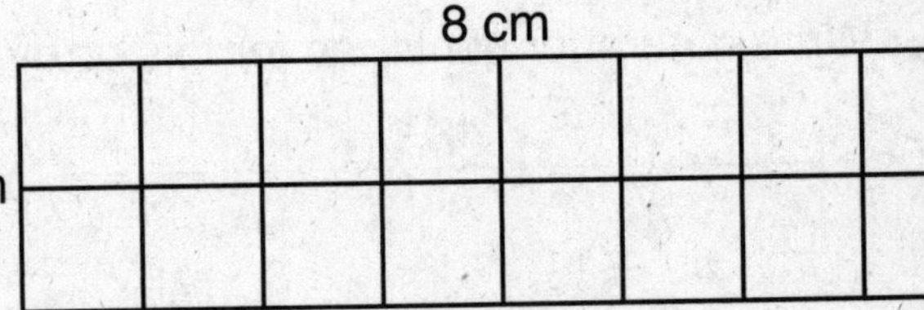

4. Someone measured the sides of the rectangle in Problem 3 using centimeters.

What is the perimeter of the rectangle? ______________

5. Write a mixed number to show how many circles are shaded.

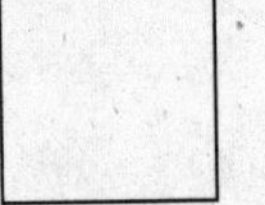

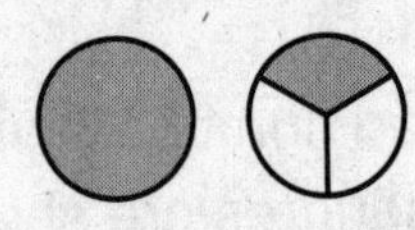

6. Put a red dot at (0, 4).
Put a blue dot at (4, 3).

4 • • • • •
3 • • • • •
2 • • • • •
1 • • • • •
0 • • • • •
  0 1 2 3 4

7. Find the answers.

$8 \div 2 =$ ______ $10 \div 2 =$ ______

$14 \div 2 =$ ______ $2 \div 2 =$ ______

$$\begin{array}{r} 284 \\ +\ 309 \\ \hline \end{array} \qquad \begin{array}{r} 57 \\ -\ 26 \\ \hline \end{array}$$

Name ______________________________

Class Fact Practice **129A**

***Saxon** Math 2 (for use with **Lesson 129**)*

Set 25: Multiplying by 4

4×3	4×9	4×4	4×1	4×6
4×8	4×0	4×5	4×7	4×2
4×6	4×4	4×9	4×1	4×8
4×3	4×7	4×2	4×5	4×0
4×7	4×1	4×4	4×8	4×3

Name ______________________________

Saxon *Math 2 (for use with* ***Lesson 129****)*

Set 25: Multiplying by 4

1. Read the answers to someone.
2. Write the answers.
3. Ask someone to correct your paper. Corrected by ______________________

$4 \times 7 =$ ___ $4 \times 0 =$ ___ $4 \times 5 =$ ___ $4 \times 9 =$ ___ $4 \times 1 =$ ___

$4 \times 4 =$ ___ $4 \times 8 =$ ___ $4 \times 2 =$ ___ $4 \times 6 =$ ___ $4 \times 3 =$ ___

$4 \times 6 =$ ___ $4 \times 4 =$ ___ $4 \times 1 =$ ___ $4 \times 8 =$ ___ $4 \times 2 =$ ___

$4 \times 0 =$ ___ $4 \times 7 =$ ___ $4 \times 5 =$ ___ $4 \times 3 =$ ___ $4 \times 9 =$ ___

$4 \times 4 =$ ___ $4 \times 8 =$ ___ $4 \times 2 =$ ___ $4 \times 7 =$ ___ $4 \times 3 =$ ___

Name ______________________

Lesson Worksheet 129A

Saxon *Math 2 (for use with* ***Lesson 129****)*

1.

Number sentence ______________________

Area = ________ square tiles

2.

Number sentence ______________________

Area = ________ square tiles

3.

Number sentence ______________________

Area = ________ square tiles

4.

Number sentence ______

Area = ______ square tiles

5.

Number sentence ______

Area = ______ square tiles

Name •

Draw a $2\frac{1}{2}$" line segment.

Date •

Measure this line segment using inches. ______ "

1. Stephanie bought a pencil for 40¢ and an eraser for 20¢. How much did she pay for the pencil and eraser?

   Number sentence ____________________

   Answer ____________________

   How much change will she get back from one dollar? ____________

2. What is the area of this rectangle?

   Area = ______ square units

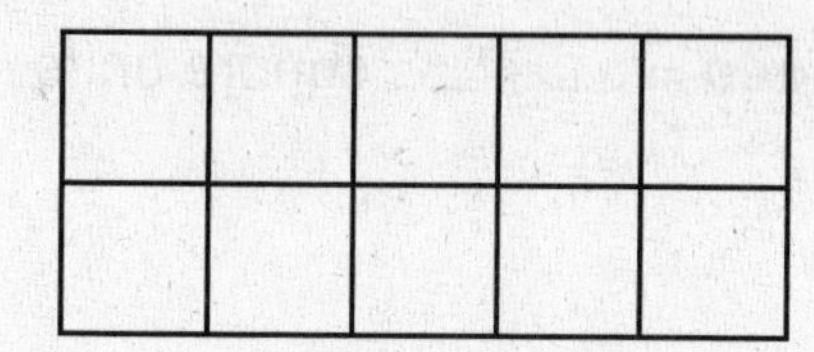

3. This is the time I get up in the morning.

   What time is it? ____________

   Write that time using words.

   ____

4. Circle all the perpendicular line segments.

5. Write a mixed number to show how many squares are shaded.

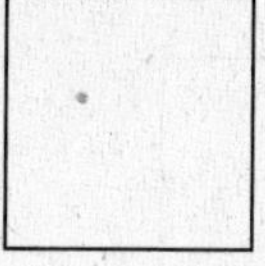 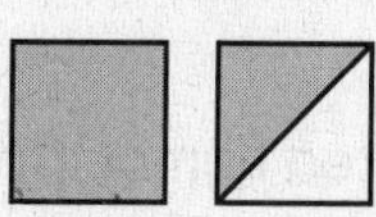

6. Write the correct comparison symbol (>, <, or =).

$16 \div 2$ ◯ 3×4 $329 + 255$ ◯ $740 - 236$

Name ______________________________

Date ______________________________

Saxon *Math 2 (for use with* ***Lesson 129****)*

1. Gladys bought a ruler for 30¢ and a marker for 40¢. How much did she pay for the ruler and marker?

Number sentence ______________________________

Answer ______________________

How much change will she get back from one dollar? ______________

2. What is the area of this rectangle?

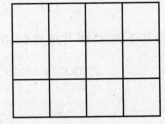

Area = ________ square units

3. This is the time I eat lunch.

What time is it? ________________

Write that time using words.

____

4. Circle all the perpendicular line segments.

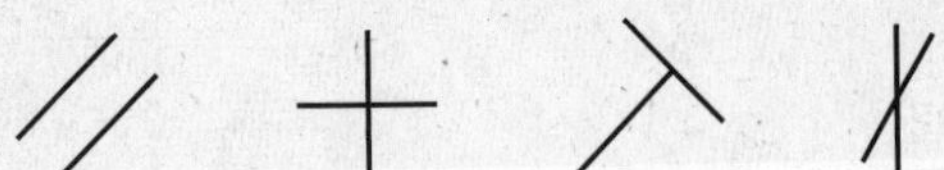

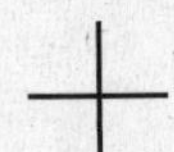

5. Write a mixed number to show how many circles are shaded.

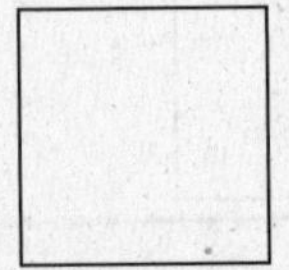

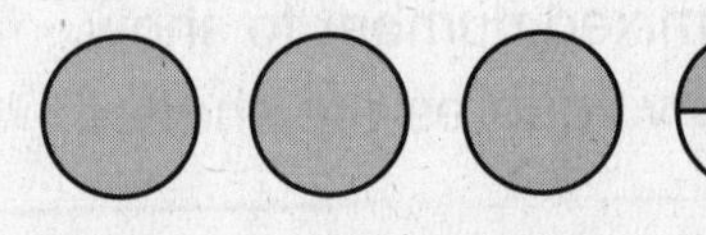

6. Write the correct comparison symbol (>, <, or =).

$18 \div 2$ ◯ 3×3

$243 + 378$ ◯ $950 - 338$

Name ______________________ Score ______

Fact Assessment 25-1

***Saxon** Math 2 (for use with **Lesson 130-1**)*

Set 25: Multiplying by 4

4×3	4×5	4×0	4×7	4×2
4×9	4×1	4×6	4×4	4×8
4×2	4×7	4×4	4×9	4×0
4×5	4×1	4×8	4×3	4×6
4×4	4×0	4×2	4×6	4×5

Name ______________________________

Fact Homework 130A

Saxon *Math 2 (for use with* ***Lesson 130-1****)*

Set 25: Multiplying by 4

Pretend you are the teacher.
Correct this paper.
If the answer is incorrect, write the correct answer next to the problem.

$4 \times 3 = 12$	$4 \times 5 = 20$	$4 \times 0 = 0$	$4 \times 7 = 26$	$4 \times 2 = 8$
$4 \times 9 = 36$	$4 \times 1 = 4$	$4 \times 6 = 28$	$4 \times 4 = 16$	$4 \times 8 = 32$
$4 \times 2 = 8$	$4 \times 7 = 28$	$4 \times 4 = 16$	$4 \times 9 = 38$	$4 \times 0 = 0$
$4 \times 5 = 20$	$4 \times 1 = 4$	$4 \times 8 = 32$	$4 \times 3 = 12$	$4 \times 6 = 24$
$4 \times 4 = 16$	$4 \times 0 = 4$	$4 \times 2 = 8$	$4 \times 6 = 28$	$4 \times 5 = 20$

Name ______________________________

Saxon Math 2 (for use with **Lesson 130-1**)

S100: 100 Subtraction Facts Corrected by ________

1	$7 - 1$	$10 - 4$	$9 - 0$	$16 - 9$	$5 - 4$	$12 - 6$	$9 - 7$	$11 - 3$	$8 - 2$	$6 - 6$
2	$9 - 4$	$6 - 3$	$11 - 6$	$10 - 2$	$6 - 1$	$12 - 8$	$2 - 0$	$9 - 3$	$7 - 2$	$6 - 5$
3	$5 - 5$	$11 - 4$	$4 - 2$	$15 - 9$	$8 - 0$	$10 - 6$	$14 - 5$	$9 - 9$	$4 - 0$	$12 - 7$
4	$9 - 5$	$17 - 9$	$8 - 4$	$13 - 8$	$9 - 2$	$11 - 5$	$15 - 6$	$5 - 1$	$8 - 5$	$16 - 8$
5	$8 - 6$	$11 - 7$	$1 - 0$	$7 - 3$	$9 - 6$	$4 - 3$	$17 - 8$	$10 - 5$	$12 - 4$	$13 - 7$
6	$8 - 3$	$16 - 7$	$10 - 3$	$4 - 1$	$6 - 2$	$13 - 5$	$7 - 0$	$14 - 9$	$11 - 2$	$10 - 8$
7	$13 - 9$	$10 - 7$	$18 - 9$	$14 - 6$	$1 - 1$	$12 - 3$	$7 - 5$	$2 - 1$	$11 - 8$	$7 - 7$
8	$2 - 2$	$12 - 5$	$3 - 1$	$15 - 7$	$10 - 1$	$6 - 0$	$13 - 4$	$5 - 2$	$9 - 8$	$3 - 0$
9	$11 - 9$	$7 - 6$	$13 - 6$	$3 - 3$	$14 - 8$	$9 - 1$	$6 - 4$	$12 - 9$	$7 - 4$	$8 - 7$
10	$4 - 4$	$15 - 8$	$3 - 2$	$5 - 0$	$5 - 3$	$8 - 8$	$14 - 7$	$10 - 9$	$0 - 0$	$8 - 1$

Name ______________________________

Lesson Worksheet 130-1

***Saxon** Math 2 (for use with **Lesson 130-1**)*

Multiples of 5

1	2	3	4	5	6	7	8	9	10
11	12	13	14	15	16	17	18	19	20
21	22	23	24	25	26	27	28	29	30
31	32	33	34	35	36	37	38	39	40
41	42	43	44	45	46	47	48	49	50
51	52	53	54	55	56	57	58	59	60
61	62	63	64	65	66	67	68	69	70
71	72	73	74	75	76	77	78	79	80
81	82	83	84	85	86	87	88	89	90
91	92	93	94	95	96	97	98	99	100

Multiples of 2

1	2	3	4	5	6	7	8	9	10
11	12	13	14	15	16	17	18	19	20
21	22	23	24	25	26	27	28	29	30
31	32	33	34	35	36	37	38	39	40
41	42	43	44	45	46	47	48	49	50
51	52	53	54	55	56	57	58	59	60
61	62	63	64	65	66	67	68	69	70
71	72	73	74	75	76	77	78	79	80
81	82	83	84	85	86	87	88	89	90
91	92	93	94	95	96	97	98	99	100

Multiples of 3

1	2	3	4	5	6	7	8	9	10
11	12	13	14	15	16	17	18	19	20
21	22	23	24	25	26	27	28	29	30
31	32	33	34	35	36	37	38	39	40
41	42	43	44	45	46	47	48	49	50
51	52	53	54	55	56	57	58	59	60
61	62	63	64	65	66	67	68	69	70
71	72	73	74	75	76	77	78	79	80
81	82	83	84	85	86	87	88	89	90
91	92	93	94	95	96	97	98	99	100

Multiples of 4

1	2	3	4	5	6	7	8	9	10
11	12	13	14	15	16	17	18	19	20
21	22	23	24	25	26	27	28	29	30
31	32	33	34	35	36	37	38	39	40
41	42	43	44	45	46	47	48	49	50
51	52	53	54	55	56	57	58	59	60
61	62	63	64	65	66	67	68	69	70
71	72	73	74	75	76	77	78	79	80
81	82	83	84	85	86	87	88	89	90
91	92	93	94	95	96	97	98	99	100

Name ______________________________

Date ______________________________

Problem-Solving Worksheet 130A

***Saxon** Math 2 (for use with **Lesson 130-1**)*

Understand	Plan	Solve	Check

Guess and Check

Taylor has a domino with a total of 6 dots. One-half of the domino has two more dots than the other half. Show what the dots on her domino look like.

How many dots are on each half of the domino? ________ and ________

Name ______________________

Understand	Plan	Solve	Check

Denise has a domino with a total of 7 dots. One half of the domino has three more dots than the other half. Show what the dots on her domino look like.

How many dots are on each half of the domino? ________ and ________

Circle the problem-solving strategies you used to solve this problem.

Act It Out

Draw a Picture

Make an Organized List

Make a Table

Use Logical Reasoning

Look for a Pattern

Guess and Check

Explain how you got your answer: ______________________

____

____

____

Name ______________________ Score ________

Fact Assessment 25-2

***Saxon** Math 2 (for use with **Lesson 130-2**)*

S100: 100 Subtraction Facts

1	$7 - 1$	$10 - 4$	$9 - 0$	$16 - 9$	$5 - 4$	$12 - 6$	$9 - 7$	$11 - 3$	$8 - 2$	$6 - 6$
2	$9 - 4$	$6 - 3$	$11 - 6$	$10 - 2$	$6 - 1$	$12 - 8$	$2 - 0$	$9 - 3$	$7 - 2$	$6 - 5$
3	$5 - 5$	$11 - 4$	$4 - 2$	$15 - 9$	$8 - 0$	$10 - 6$	$14 - 5$	$9 - 9$	$4 - 0$	$12 - 7$
4	$9 - 5$	$17 - 9$	$8 - 4$	$13 - 8$	$9 - 2$	$11 - 5$	$15 - 6$	$5 - 1$	$8 - 5$	$16 - 8$
5	$8 - 6$	$11 - 7$	$1 - 0$	$7 - 3$	$9 - 6$	$4 - 3$	$17 - 8$	$10 - 5$	$12 - 4$	$13 - 7$
6	$8 - 3$	$16 - 7$	$10 - 3$	$4 - 1$	$6 - 2$	$13 - 5$	$7 - 0$	$14 - 9$	$11 - 2$	$10 - 8$
7	$13 - 9$	$10 - 7$	$18 - 9$	$14 - 6$	$1 - 1$	$12 - 3$	$7 - 5$	$2 - 1$	$11 - 8$	$7 - 7$
8	$2 - 2$	$12 - 5$	$3 - 1$	$15 - 7$	$10 - 1$	$6 - 0$	$13 - 4$	$5 - 2$	$9 - 8$	$3 - 0$
9	$11 - 9$	$7 - 6$	$13 - 6$	$3 - 3$	$14 - 8$	$9 - 1$	$6 - 4$	$12 - 9$	$7 - 4$	$8 - 7$
0	$4 - 4$	$15 - 8$	$3 - 2$	$5 - 0$	$5 - 3$	$8 - 8$	$14 - 7$	$10 - 9$	$0 - 0$	$8 - 1$

Name ______________________

Date ______________________

Written Assessment 25

***Saxon** Math 2 (for use with **Lesson 130-2**)*

1. There are 5 tables in Room A. There are 3 books on each table. Draw a picture of the books on the tables. How many books are on the tables altogether?

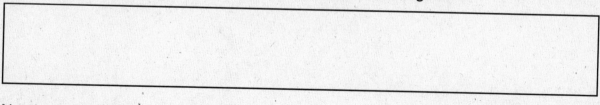

Number sentence ______________________

Answer ______________________

2. Label this array.
Write a number sentence for this array.

____

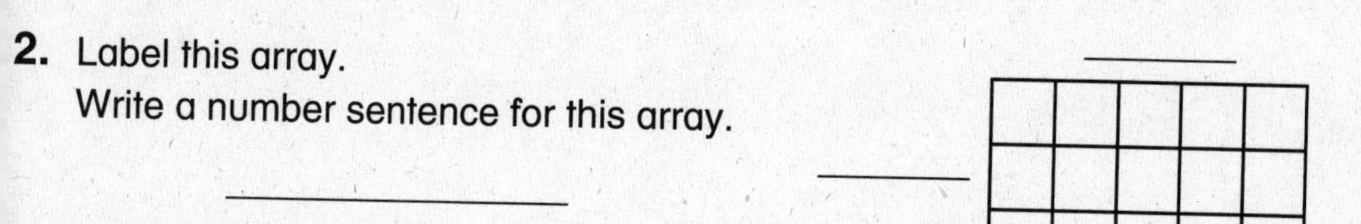

3. Use a crayon to trace an example of perpendicular line segments on this paper.
Where do you see perpendicular line segments in the classroom?

____

4. There are 3 green, 1 blue, and 8 red candies in a bag.
If you pick a candy without looking,
which color will you most likely pick? ______________________

Why? ______________________

5. Write a mixed number to show how many squares are shaded.

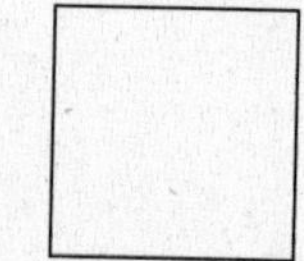

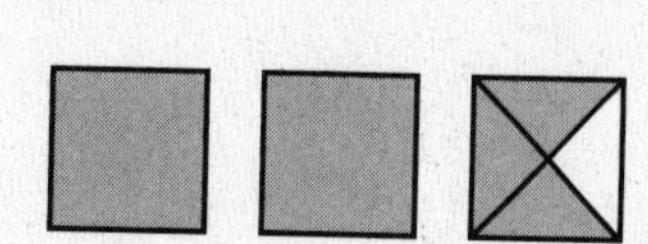

6. Find the answers.

$$\begin{array}{r} 423 \\ +\ 294 \\ \hline \end{array} \qquad \begin{array}{r} \$9.26 \\ -\ 5.65 \\ \hline \end{array} \qquad \begin{array}{r} 382 \\ -\ 124 \\ \hline \end{array} \qquad \begin{array}{r} \$6.27 \\ +\ 1.95 \\ \hline \end{array}$$

Name ______________________________

Lesson Worksheet 130-2

***Saxon** Math 2 (for use with **Lesson 130-2**)*

Multiplication Table

X	0	1	2	3	4	5	6	7	8	9
0										
1										
2										
3										
4										
5										
6										
7										
8										
9										

Name ______________________________

Saxon *Math 2 (for use with* ***Lesson 131****)*

0, 3, 4, 5, 6, 8, 9, 10, 12, 14, 15, 16, 18,
20, 21, 24, 25, 27, 28, 30, 32, 35, 36, 40, 45

Name ______________________________

Saxon Math 2 (for use with Lesson 131)

Set 26: Multiplying by 2, 3, 4, and 5 Corrected by ________

Write the answers.

0×2	1×2	2×2	3×2	4×2
5×2	6×2	7×2	8×2	9×2
0×3	1×3	2×3	3×3	4×3
5×3	6×3	7×3	8×3	9×3
0×4	1×4	2×4	3×4	4×4
5×4	6×4	7×4	8×4	9×4
0×5	1×5	2×5	3×5	4×5
5×5	6×5	7×5	8×5	9×5

Name • Draw a 7-cm line segment.

Guided Class Practice 131A

Saxon Math 2 (for use with Lesson 131)

Date •——————————————————•

Measure this line segment using centimeters. ________ cm

1. Mrs. Voulgaris bought a hat and a cup at the school fair. How much money did she spend?

Hat	$3.78
Book	$2.95
Cup	$1.61

Workspace

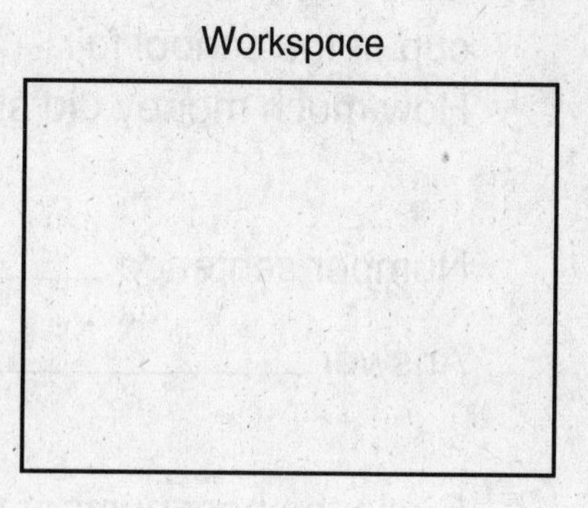

Number sentence ______________________________

Answer ____________________________

2. Circle the best number sentence to use to estimate the sum of 119 and 62.

120 + 60 = 180   100 + 70 = 170

110 + 60 = 170   120 + 70 = 190

3. Where is the ✶ on this coordinate graph?

____

Put a red dot at (1, 0).

4 • • • • •
3 • • • • •
2 • • • ✶ •
1 • • • • •
0 • • • • •
0 1 2 3 4

4. The school talent show started at one o'clock. It lasted one hour and fifteen minutes. Show the time it ended on both clocks.

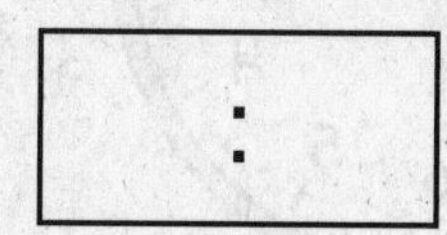

5. Fill in the correct comparison symbol (>, <, or =).

3 × 4 ◯ 6 × 2   16 ÷ 2 ◯ 15 − 8

________ ________   ________ ________

6. Find the differences.

84 − 47   169 − 86 $5.29 − $2.45

Name ______________________

Date ______________________

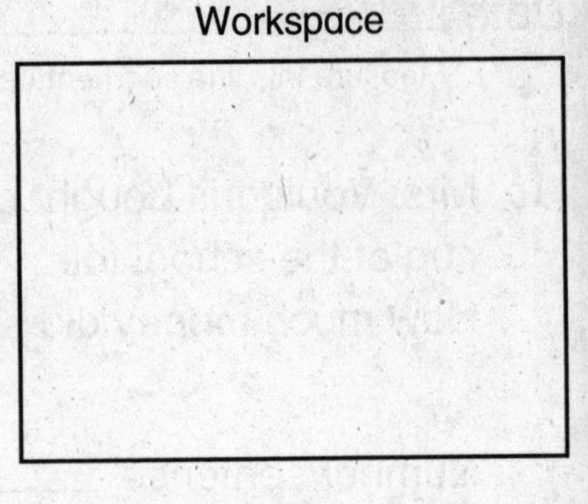

1. Miss Keyes bought a book and a cup at the school fair. How much money did she spend?

Hat	$3.78
Book	$2.95
Cup	$1.61

Number sentence ______________________

Answer ______________________

2. Circle the best number sentence to use to estimate the sum of 127 and 54.

130 + 50 = 180 130 + 60 = 190

120 + 50 = 170 120 + 60 = 180

3. Where is the ★ on this coordinate graph?

______

Put a red dot at (3, 0).

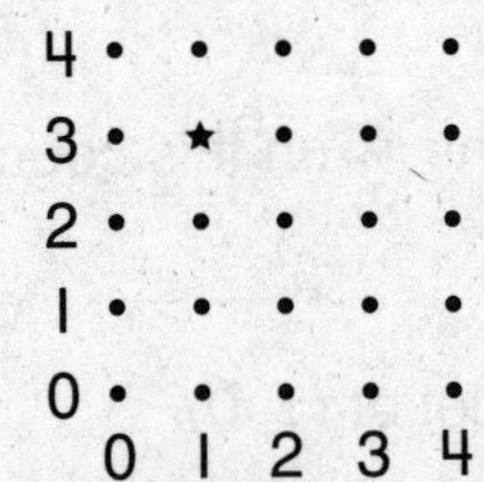

4. The movie started at three o'clock. It lasted one hour and thirty minutes. Show the time it ended on both clocks.

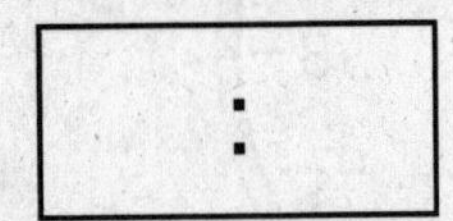

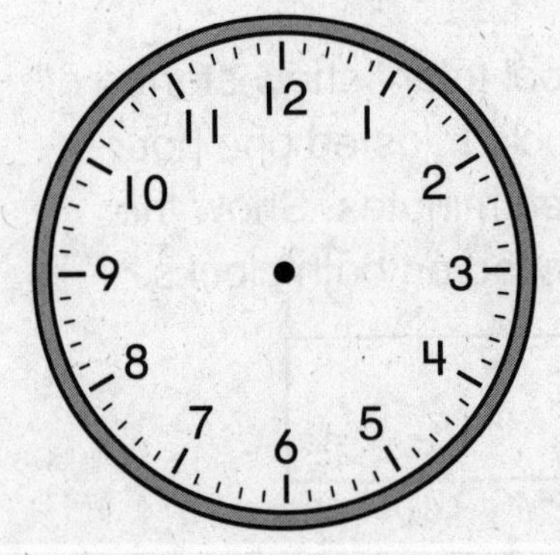

5. Fill in the correct comparison symbol (>, <, or =).

4 × 4 ◯ 3 × 5   12 ÷ 2 ◯ 15 − 9

______ ______   ______ ______

6. Find the differences.

78 − 39   190 − 73 $4.73 − $1.65

Name ______________________________

Class Fact Practice 132A

***Saxon** Math 2 (for use with **Lesson 132**)*

Set 26: Multiplying by 2, 3, 4, and 5

2×5　　5×6　　3×0　　4×1　　2×2

5×3　　2×8　　4×6　　3×3　　5×9

3×6　　4×2　　5×4　　2×1　　3×9

2×3　　4×4　　5×1　　3×4　　5×7

4×9　　5×8　　2×0　　4×7　　3×1

2×7　　4×3　　3×8　　5×0　　2×6

4×5　　3×7　　2×4　　4×8　　3×5

5×2　　2×9　　4×0　　3×2　　5×5

$6 \times 3 = 18$　　$8 \times 3 = 24$　　$6 \times 4 = 24$　　$8 \times 4 = 32$

$7 \times 3 = 21$　　$9 \times 3 = 27$　　$7 \times 4 = 28$　　$9 \times 4 = 36$

Name ______________________

Fact Homework 132B

***Saxon** Math 2 (for use with **Lesson 132**)*

Set 26: Multiplying by 2, 3, 4, and 5

Corrected by ________

2×6	5×0	3×8	4×3	2×7
3×1	4×7	2×0	5×8	4×9
5×7	3×4	5×1	4×4	2×3
3×9	2×1	5×4	4×2	3×6
5×9	3×3	4×6	2×8	5×3
2×2	4×1	3×0	5×6	2×5
5×5	3×2	4×0	2×9	5×2
3×5	4×8	2×4	3×7	4×5

$6 \times 3 = 18$	$8 \times 3 = 24$	$6 \times 4 = 24$	$8 \times 4 = 32$
$7 \times 3 = 21$	$9 \times 3 = 27$	$7 \times 4 = 28$	$9 \times 4 = 36$

Name ______________________

Date ______________________

Guided Class Practice 132A

Saxon Math 2 (for use with Lesson 132)

1. Quinton has 9 baseball cards. Curtis has twice as many baseball cards as Quinton. How many baseball cards does Curtis have?

Answer ______________________

2. Stickers cost 4¢ each at the school store. Use the table to show the cost of 10 stickers.

Stickers	1	2	3	4	5	6				
Cost	4¢	8¢	12¢							

How many stickers can you buy with 20¢? ______________

3. This is the time I eat breakfast.

What time is it? ______________

Write the time using words.

____

4. Double the value of these coupons.

20¢ ______________ 35¢ ______________

5. The area of this rectangle is the number of square centimeter tiles you will need to cover it.

☐ = 1 square centimeter

What is the area of the rectangle?

Number sentence ______________________ Area = ________ square cm

6. Find the answers.

54 + 36 + 27 60 − 7 129 + 85 $3.25 − $1.50

Name ______________________________

Date ______________________________

Saxon Math 2 (for use with Lesson 132)

1. Ariana has 7 stuffed animals. Her sister has twice as many stuffed animals. How many stuffed animals does Ariana's sister have?

Answer ______________________

2. Pencils cost 5¢ each at the school store. Use the table to show the cost of 10 pencils.

Pencils	1	2	3	4	5	6				
Cost	5¢	10¢	15¢							

How many pencils can you buy with 40¢? ______________

3. This is the time I go to bed.

What time is it? ______________

Write the time using words.

____

4. Double the value of these coupons.

15¢ ______________

40¢ ______________

5. The area of this rectangle is the number of square centimeter tiles you will need to cover it.

☐ = 1 square centimeter

6 cm

2 cm

What is the area of the rectangle?

Number sentence ______________________ Area = ______ square cm

6. Find the answers.

48 + 63 + 12

50 − 8

164 + 37

$5.75 − $2.90

+ ______

− ______

+ ______

− ______

Name ______________________________

Saxon Math 2 (for use with Lesson 133)

Set 26: Multiplying by 2, 3, 4, and 5

Corrected by ______

4×2	5×4	2×1	3×9	2×3
4×4	5×1	3×4	5×7	4×9
5×8	2×0	4×7	3×1	2×7
4×3	3×8	5×0	2×6	4×5
3×7	2×4	4×8	3×5	5×2
2×9	4×0	3×2	5×5	2×5
5×6	3×0	4×1	2×2	5×3
2×8	4×6	3×3	5×9	3×6

$6 \times 3 = 18$	$8 \times 3 = 24$	$6 \times 4 = 24$	$8 \times 4 = 32$
$7 \times 3 = 21$	$9 \times 3 = 27$	$7 \times 4 = 28$	$9 \times 4 = 36$

Name ______________________

6 2 groups of ____ 3 groups of ____	**8** 2 groups of ____ 4 groups of ____
10 2 groups of ____ 5 groups of ____	**15** 3 groups of ____ 5 groups of ____
12 2 groups of ____ 3 groups of ____ 4 groups of ____ 6 groups of ____	**20** 2 groups of ____ 4 groups of ____ 5 groups of ____ 10 groups of ____

Name ______________________

Date ______________________

Guided Class Practice 133A

Saxon Math 2 (for use with Lesson 133)

1. Mrs. Velardi's class has 12 markers. Each child will need 3 markers. Draw a picture to show how many children can have markers.

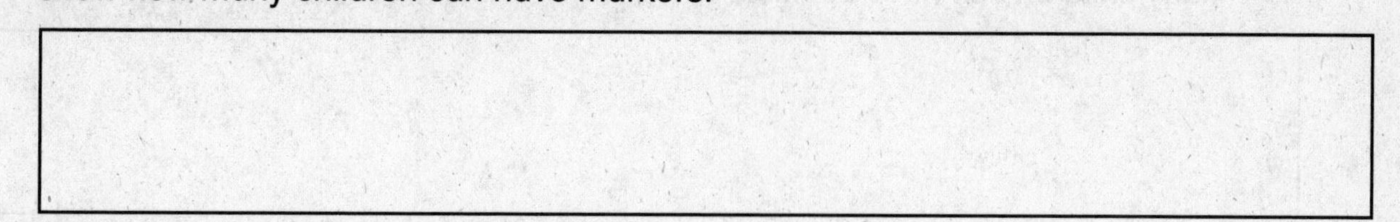

How many children can have markers? ______________

2. Circle the best number sentence to use to estimate the sum of 47 and 39.

40 + 40 = 80 40 + 30 = 70 50 + 30 = 80 50 + 40 = 90

3. Where is the ☺ on this coordinate graph?

(,)

Put a blue dot at (0, 3).

4	•	•	•	•	•
3	•	•	•	•	•
2	•	•	•	•	•
1	•	•	•	•	☺
0	•	•	•	•	•
	0	1	2	3	4

4. James, Nicole, Clem, and Crystal will share a dozen muffins equally. Draw the muffins on the children's plates.

How many muffins will each child have?

____

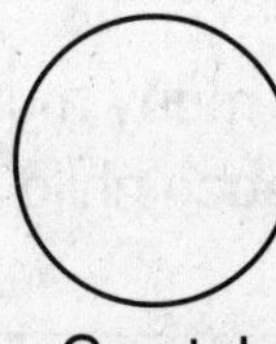

James Nicole Clem Crystal

5. One of the digits [6], [7], or [8] belongs in each of the problems below. Fill in the correct digit in each problem.

```
  4 1        3 □        7 5
+ 2 □      + 5 2      + 1 □
-----      -----      -----
  6 8        9 0        9 1
```

6. Fill in the correct comparison symbol (>, <, or =).

6 × 2 ◯ 4 × 3 14 ÷ 2 ◯ 15 − 7 6 × 5 ◯ 7 × 4

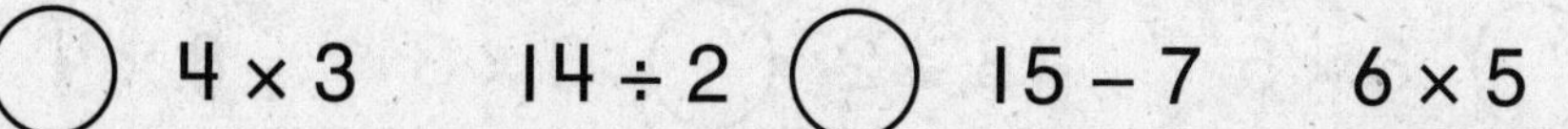

Name ______________________________

Date ______________________________

Saxon Math 2 (for use with **Lesson 133**)

1. Eddie has 10 cookies. He wants to give each child 2 cookies. Draw a picture to show how many children can have cookies.

How many children can have cookies? __________

2. Circle the best number sentence to use to estimate the sum of 28 and 53.

$20 + 50 = 70$ $30 + 50 = 80$ $20 + 60 = 80$ $30 + 60 = 90$

3. Where is the ☺ on this coordinate graph?

(,)
____

Put a blue dot at (0, 4).

4 • • ☺ • •
3 • • • • •
2 • • • • •
1 • • • • •
0 • • • • •
0 1 2 3 4

4. Rodney, Jason, and Luis will share a dozen muffins equally. Draw the muffins on the children's plates.

How many muffins will each child have?

____

Rodney Jason Luis

5. One of the digits 2, 5, or 9 belongs in each of the problems below. Fill in the correct digit in each problem.

2 3 + 3 ☐ 5 8	3 ☐ + 4 1 8 0	6 9 + 1 ☐ 8 1

6. Fill in the correct comparison symbol (>, <, or =).

7×2  3×5 $18 \div 2$  3×3 8×4 ◯ 7×5

______ ______ ______ ______ ______ ______

Name ____________________

Class Fact Practice 134A

***Saxon** Math 2 (for use with **Lesson 134**)*

Set 26: Multiplying by 2, 3, 4, and 5

3×1	2×5	5×8	4×2	2×3
5×2	2×0	4×4	3×5	5×6
4×9	3×2	2×7	5×3	3×8
5×5	3×6	4×3	2×1	5×9
2×4	4×0	3×3	5×4	4×7
3×9	5×1	2×8	4×5	3×0
2×2	5×7	3×4	4×1	2×6
4×8	5×0	2×9	3×7	4×6

Name ______________________

Saxon Math 2 (for use with **Lesson 134**)

Set 26: Multiplying by 2, 3, 4, and 5

Corrected by ______

3×0	4×5	2×8	5×1	3×9
2×6	4×1	3×4	5×7	2×2
4×6	3×7	2×9	5×0	4×8
2×3	4×2	5×8	2×5	3×1
5×6	3×5	4×4	2×0	5×2
3×8	5×3	2×7	3×2	4×9
5×9	2×1	4×3	3×6	5×5
4×7	5×4	3×3	4×0	2×4

Name ______________________________

Saxon Math 2 (for use with **Lesson 134**)

Domino Sums

0 1 2 3 4 5 6 7 8 9 10 11 12

Name ______________________

Date ______________________

Guided Class Practice 134A

***Saxon** Math 2 (for use with **Lesson 134**)*

1. Lou's dog eats 3 dog biscuits a day. How many dog biscuits will the dog eat in 7 days?

Number sentence ______________________

Answer ______________________

2. There are twice as many girls in Mrs. York's class as there are boys.

Are there more girls or boys in her class? ____________

There are 10 girls in Mrs. York's class.

How many boys are in her class? ____________

3. Mrs. Dunleavy had 8 cookies. She gave each of her children 2 cookies. Draw a picture to show the cookies.

How many children does Mrs. Dunleavy have? ________

4. The cost of the pencil is 60¢.
How much change will you receive from $1.00? ____________

5. Stephen can buy 2 folders for 25¢. Use the table to show the cost of 14 folders.

Folders	2	4	6	8			14
Cost	25¢	50¢					

How many folders can Stephen buy with $1.00? ____________

6. Write the quotients.

14 ÷ 2 = ______ 8 ÷ 2 = ______ 18 ÷ 2 = ______

7. Find the answers.

$$\begin{array}{r} \$2.93 \\ +\ 1.42 \\ \hline \end{array} \qquad \begin{array}{r} \$5.95 \\ -\ 1.78 \\ \hline \end{array} \qquad \begin{array}{r} \$6.29 \\ -\ 3.49 \\ \hline \end{array}$$

Name ______________________________

Date ______________________________

***Saxon** Math 2 (for use with **Lesson 134**)*

1. Pam's dog eats 2 dog biscuits a day. How many dog biscuits will the dog eat in 7 days?

Number sentence ______________________________

Answer ______________________________

2. There are twice as many boys in Mrs. Rivera's class as there are girls.

Are there more girls or boys in her class? ____________________

There are 12 boys in Mrs. Rivera's class.

How many girls are in her class? ____________________

3. Mr. VanWinkle had 12 cookies. He gave each of his children 4 cookies. Draw a picture to show the cookies.

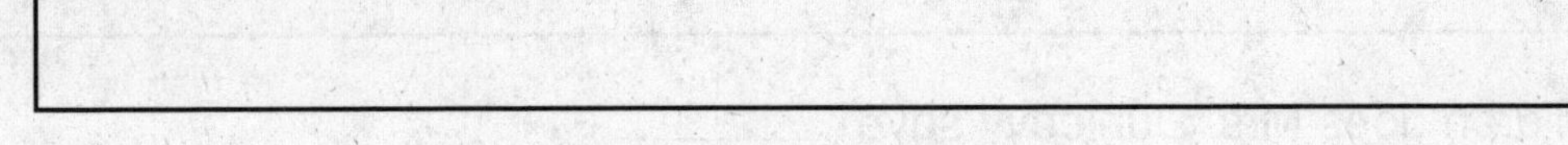

How many children does Mr. VanWinkle have? ________

4. The cost of the ruler is 70¢.
How much change will you receive from $1.00? ____________________

5. Courtney can buy 3 pencils for 10¢. Use the table to show the cost of 21 pencils.

Pencils	3	6	9	12			21
Cost	10¢	20¢					

How many pencils can Courtney buy with 50¢? ____________________

6. Write the quotients.

10 ÷ 2 = ________ 6 ÷ 2 = ________ 12 ÷ 2 = ________

7. Find the answers.

$$\begin{array}{r} \$4.65 \\ +\ \ 3.74 \\ \hline \end{array} \qquad \begin{array}{r} \$7.28 \\ -\ \ 5.09 \\ \hline \end{array} \qquad \begin{array}{r} \$5.25 \\ -\ \ 2.75 \\ \hline \end{array}$$

Name ______________________ Score ________

Fact Assessment 26

***Saxon** Math 2 (for use with **Lesson 135**)*

Set 26: Multiplying by 2, 3, 4, and 5

5×2	2×1	3×3	4×0	5×3
2×7	3×0	4×3	2×4	5×7
3×5	2×2	5×4	4×4	2×6
5×9	3×1	2×9	4×2	3×4
2×3	5×6	4×5	2×8	5×0
4×1	2×5	5×8	3×2	5×5

Name ______________________

Date ______________________

Written Assessment 26

***Saxon** Math 2 (for use with **Lesson 135**)*

1. Maria bought a puzzle and a box of crayons. How much money did she spend?

Puzzle	$2.55
Markers	$3.47
Crayons	$1.29

Workspace

Number sentence ______________________

Answer ______________________

2. How many children chose pink? ________

How many children chose blue? ________

Shade the graph to show that 9 children chose green.

How many more children chose blue than chose pink? ________

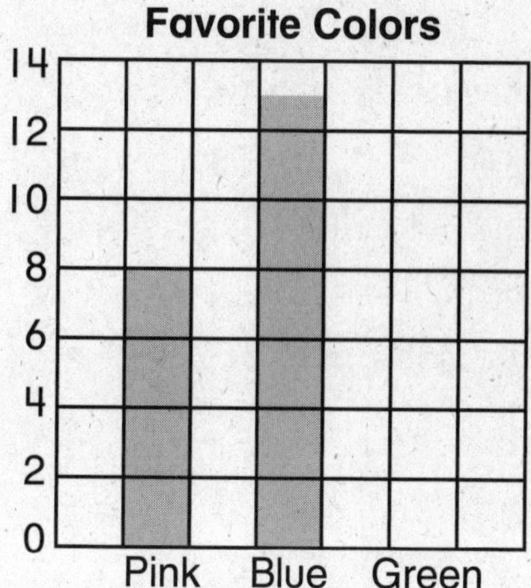

3. Where is the ✶ on this graph?

(___ , ___)

Put a red dot at (2, 4).

4. Circle the clock that shows quarter to seven.

5. Circle the best number sentence to use to estimate the sum of 53 and 27.

50 + 20 = 70 60 + 20 = 80 50 + 30 = 80 60 + 30 = 90

6. Fill in the correct comparison symbol (>, <, or =).

3 × 4  6 × 2 16 ÷ 2  16 − 9 5 × 3  4 × 4

________ ________ ________ ________ ________ ________

Name ______________________

***Saxon** Math 2 (for use with **Lesson 135**)*

Domino Probability Experiment

0	1	2	3	4	5	6	7	8	9	10	11	12

Name ______________________

Date ______________________

Guided Class Practice 135A

*Saxon Math 2 (for use with **Lesson 135**)*

1. Curtis wants to buy 2 markers and 4 erasers. How much money will he need?

	Single	Package of 2	Package of 4
Pencils	15¢	25¢	40¢
Markers	30¢	50¢	85¢
Erasers	10¢	15¢	25¢

Number sentence ______________________

Answer ______________________

2. Use the bar graph to answer the questions.

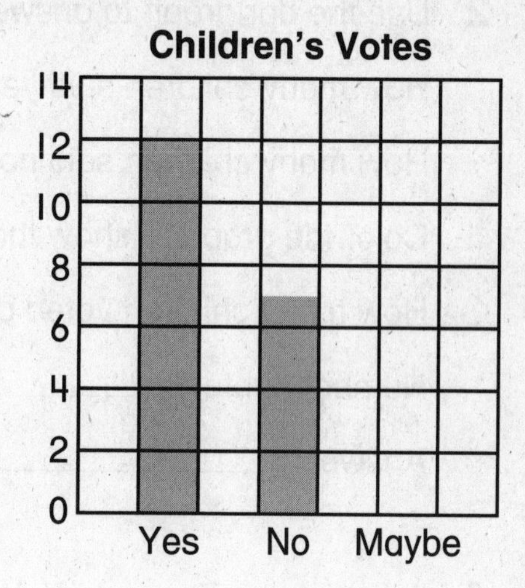

How many children said yes? ________

How many children said no? ________

Color the graph to show that 5 children said maybe.

How many children voted altogether?

Number sentence ______________________

Answer ______________________

3. Annette, Ebony, and Donna will share 15 markers. Draw the markers on the children's desks. How many markers will each child receive?

Answer ______________________

4. Write four hundred twenty using digits. ________

Write this number in expanded form. ______________________

5. One of the digits 2, 5, or 9 belongs in each of the problems below. Fill in the correct digit in each problem.

```
  2 6 4          8 □          2 1 9
+   4 □      +   1 9       + 4 3 □
-------      -------       -------
  3 0 9      1 0 1          6 5 8
```


Name ______________________________

Date ______________________________

Saxon Math 2 (for use with Lesson 135)

1. Judy wants to buy 2 pencils and 2 markers.
How much money will she need?

Number sentence ______________________

Answer ______________________

	Single	Package of 2	Package of 4
Pencils	15¢	25¢	40¢
Markers	30¢	50¢	85¢
Erasers	10¢	15¢	25¢

2 Use the bar graph to answer the questions.

How many children said yes? ________

How many children said no? ________

Color the graph to show that 7 children said maybe.

How many children voted altogether?

Number sentence ______________________

Answer ______________________

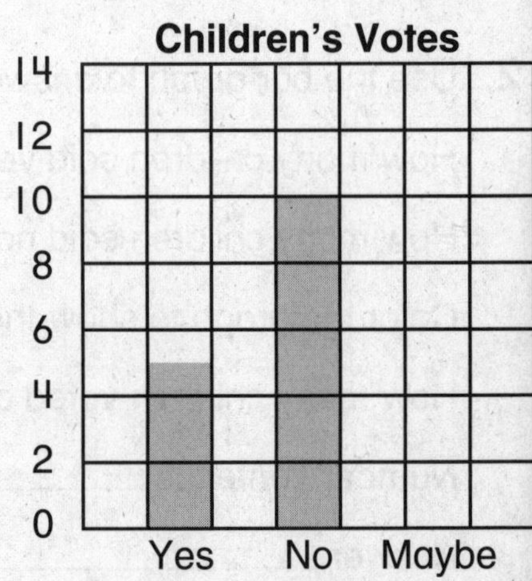

3. Alan, Lance, Ted, and Kyle will share 20 markers.
Draw the markers on the children's desks.
How many markers will each child receive?

Answer ______________________

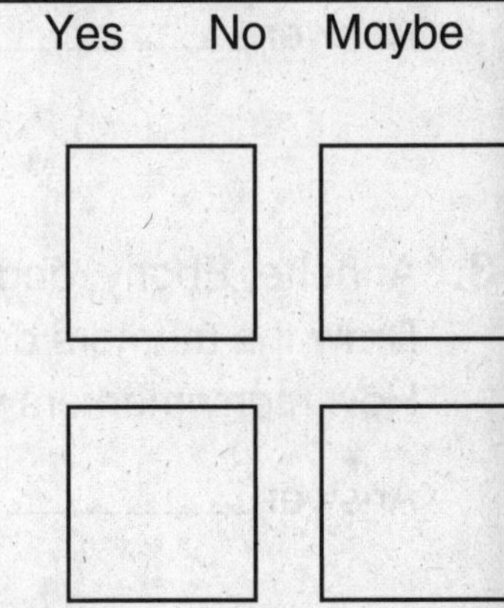

4. Write five hundred seven using digits. ________

Write this number in expanded form. ______________________

5. One of the digits [3], [4], or [7] belongs in each of the problems below.
Fill in the correct digit in each problem.

$$\begin{array}{r} 42\square \\ +\ 82 \\ \hline 506 \end{array} \qquad \begin{array}{r} 9\square \\ +\ 36 \\ \hline 133 \end{array} \qquad \begin{array}{r} 427 \\ +\ 35\square \\ \hline 780 \end{array}$$